AS THE WORLD BURNS

50 SIMPLE THINGS YOU CAN DO TO STAY IN DENIAL

AS THE WORLD BURNS

50 SIMPLE THINGS YOU CAN DO TO STAY IN DENIAL

A GRAPHIC NOVEL

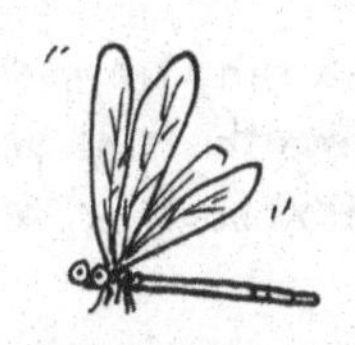

DERRICK JENSEN

AND

STEPHANIE McMILLAN

SEVEN STORIES PRESS

NEW YORK

A Seven Stories Press First Edition

Seven Stories Press
140 Watts Street
New York, NY 10013
http://www.sevenstories.com

College professors may order examination copies of Seven Stories Press titles for a free six-month trial period. To order, visit http://www.sevenstories.com/textbook or send a fax on school letterhead to (212) 226-1411.

Library of Congress Cataloging-in-Publication Data
Jensen, Derrick, 1960-
As the world burns: 50 simple things you can do to stay in denial/ by Derrick Jensen; illustrated by Stephanie McMillan.—1st ed.
p.cm.
ISBN 978-1-58322-777-0 (pbk.)
1. Environmental policy-United States-Comic books, strips, etc. 2. American wit and humor, Pictorial. I. McMillan, Stephanie. II. Title.

GE 180.J46 2007
363.7-dc22
2007036959

Printed in the USA.

9 8 7 6

TO OUR MOTHERS

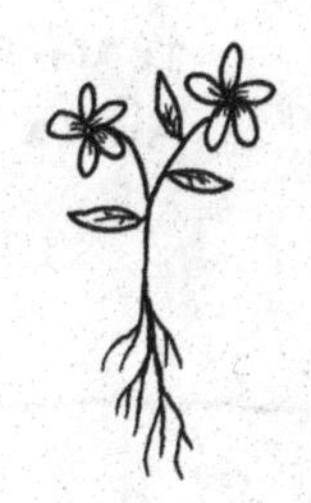

Did you know that if you recycle a single aluminum can, you can save enough energy to power a television set for three hours?

Oh my GOD! That is SO HELPFUL!!

You mean big industries can make lots of money by building dams for aluminum smelting, choking the life out of rivers and killing the fish, plus tearing up great swaths of Africa for mining bauxite?

WOW!

Big money for builders, miners and manufacturers!

Fabulous!

Then countless people can acquire Alzheimer's Disease from ingesting the toxic metal!
Drinking the soda from the cans will help millions enjoy diabetes, obesity and malnutrition!

And then they'll add the cans to millions of other used cans and melt them down using coal-burning plants that spew toxins into the air,
and then use all that metal to make more cans, using still more energy.

So you can drink more soda.

And the melting and re-forming will take somewhat less energy than making all totally new cans.
So as a FABULOUS BONUS...

You can watch three episodes of "Survivor," or even Six "Brady Bunch" reruns with the energy that's saved.
Therefore using the same amount of energy as before, but gaining three hours of brainwashing! What a brilliant scheme!

You're a real party-pooper, you know that?

We see this so often.
50 SIMPLE THINGS YOU CAN DO TO SAVE THE EARTH!
CAN STREAMING VIDEO STOP GLOBAL WARMING?
LET TIKI THE PENGUIN GUIDE YOU! FIND OUT HOW KIDS CAN STOP GLOBAL WARMING!
YAHOO CELEBRATES 35 YEARS OF EARTH DAY WITH "10 SIMPLE WAYS YOU CAN SAVE THE WORLD IN A DAY!"
GREEN YOUR EVERYDAY ONLINE SHOPPING AND STOP GLOBAL WARMING!

Wouldn't it be wonderful if life were this simple, the problems we face so easily solvable? Every cell in my body wants for recycling to save the day, wants for shorter showers to save enough water for the rivers to run free.
But they won't. You know that. Fish and turtles and beavers and frogs and bears know that. Everybody knows that.
Sometimes we forget.

I just saw a great movie about how the world is heating up, and about what we can do to stop it.
Finally the issue is being addressed? It's about damn time.
One thing we can do is to change our light bulbs.
Light bulbs.
Light bulbs?
Replacing one regular light bulb with a compact flourescent saves 150 pounds of carbon dioxide per year!
That's their solution?!
Oh, there's so much more!
I can hardly wait.

Oooh, fox-ay!

There's a layer of reality not perceived by our five senses, that connects us to everything.
You only have five?

What we're experiencing is a result of what's going on within ourselves.
We're the solution, as well as the problem, as well as the creators of the problem.
Meditation is key. The infinite source of all reality will solve the Earth's dilemma.

This will change the world.
I invite you to try it with me.

Before me peaceful.
Behind me peaceful.
Above me peaceful.
Below me ...

Ready to stop global warming?
I'm so excited!
Yeah, let's stop the burning of oil and coal!
Me too! All we need to do is dismantle the industrial economy!

So... we already talked about changing light bulbs.
You think I'm kidding.

Don't be Silly.

The movie said we should drive less. Walk, bike, carpool or take mass transit more often. You'll save one pound of carbon dioxide for every mile you don't drive.
Which means you put out one pound for every mile you do...

Recycle more: you can save 2,400 pounds of carbon dioxide per year by recycling just half of your household waste.
Which means...

Just stop it. Stop it. Stop it. Let me continue.

Okay.

You don't like my list.

Your list is fine.
It's just.... just...

Just go on. Please.

Okay! You're really going to like this one! It's really important, and you don't have to change your lifestyle at all!
Well, thank goodness for that!

Check your tires: keeping your tires inflated properly can improve gas mileage by more than 3%. Every gallon of gasoline saved keeps 20 pounds of carbon dioxide out of the atmosphere.
Which means that every gallon burned...

Why do you do this?
What?
Ruin everything. That's all you do.
Please, go on.

Use less hot water.
Will that stop global warming?

Oh yes! It takes a lot of energy to heat water.
You can install a low flow showerhead to save 350 pounds of CO2 per year, and wash your clothes in cold or warm water to save 500 pounds per year.

Or you could knock out the water supply to a factory. That would save a lot of energy, and water too!

That's it exactly! You can save 1,200 pounds of carbon dioxide if you cut down your garbage by 10 percent!

Oh, and you'll love this one, since you love nature. Plant a tree!!
A single tree will absorb one ton of CO_2 during its lifetime.
How about if we stop someone from cutting down a tree?

I suppose that would work. But wouldn't it be hard?
I'd just rather plant a little tree all my own and watch it grow and grow and grow.
Wouldn't that be fun?

We're almost there.
I can't wait! This planet is supposed to be just covered with living things!
Oh, don't make me hungry!
I know, I can't stop dreaming about it!
We'll be so happy there. We'll be able to raise our family in peace, security and abundance.
We'll reproduce like little bunnybots until every inch of the universe is blessed with replicas of our superior selves.
Just like the Great Mechanical One ordered us to do so long ago: "Go forth and be everywhere."
Mmm... I can't wait!
Hey, cut that out! I'm trying to drive!

we will go quietly, meekly, to the end of the world, if only you allow us to believe that buying low energy light bulbs will save us.

I did some math with that global warming list.*
Let's pretend everybody in the United States does everything on that list.
Wouldn't that be wonderful?
And let's pretend they change not one light bulb but every light bulb in their homes.
That would make me so happy!
You and the light bulb manufacturers both.
That would save about 400 million tons of carbon dioxide.
Yippee!! I told you this would save the world!
And that's only one item on the list!
Next, people in the United States drive about 2.3 trillion miles per year.
If we cut that in half we save about 575 million tons of carbon dioxide.
You do like my list! Oh, you do, you do!
* Math by Tim Bennet, Sally Erickson and Dan Combs.

Recycling half of every American's household waste would save about 125 million tons.
Inflating tires and thus increasing gas mileage would save about 85 million tons.
Inflate the tires, yes!

Low flow showerheads and washing clothes in cooler water would save 105 million tons.
I could do that! So could everyone else!
Cutting down on garbage by ten percent reduces emissions by 65 million tons.
And I don't even mind that we counted that one twice! We have so much to spare!
Adjusting the thermostat will save about 105 million tons.
Lowering it two degrees in the summer might be hard. It's so hot out! But to save the world I would do that, wouldn't you?
If every person in the United States planted a tree, that would absorb seven million tons of carbon dioxide a year.
I can add it all up! It makes a total of about 1.5 billion tons of carbon dioxide a year!

We did it! We saved the planet!

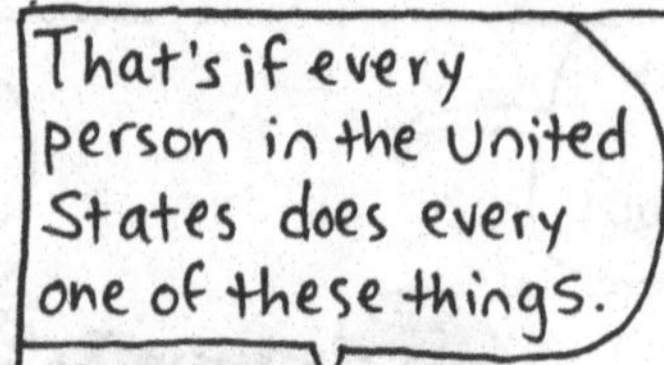
That's if every person in the United States does every one of these things.

But they will! If we just tell them. This isn't so hard! We can do it!

There's just one thing. Total carbon emissions for the United States is 7.1 billion tons.
If every man, woman and child did all of the things on the list from the movie ~ and you know there is precisely zero chance that every man, woman and child in the United States will do this...

I don't like where this is going.

That would only be about a 21 percent reduction in carbon emissions. And since total carbon emissions go up about two percent per year, that whole reduction would disappear in about...

sigh.

What's worse is that the scientific consensus is that if we are to avert disaster — or really to avert further disaster, since the oceans are already being killed, the coral reefs are dying, deserts are expanding, cold water fish are in trouble, amphibians...
I get the picture.
We need to reduce carbon emissions by 80 percent, and we need to do it now.
You don't want to plant a tree?
What will it take to reduce carbon emissions by 75 percent?
I don't think you like me.
No, I like you a lot.
Then why do you shoot down everything I say?
Want to help me change some light bulbs? I just bought them, and they didn't have any packaging.

Are we going to have any problems with the dominant life-forms there? We probably have to annihilate them all like last time, right?
On the contrary, the Great Mechanical One gave us such an easy planet this time! We probably won't have to kill a single one!

According to our guidebook, these particular creatures will do anything for gold.
Don't lie!

No, it's true! The experiment was repeated dozens of times: give them gold and they don't care what you take from them. Trees, animals, water, soil... they're so eager to trade these things away, they practically hand you a knife and fork!

But... that's so gross! Gold is our waste product! That's totally insane and disgusting! What do they do with it?
As far as our researchers could determine, absolutely nothing. They can't even get a buzz off it. They just love it better than anything else, for reasons we'll probably never understand.
Lucky us!!

People talk all the time about how the world is running out of water.
Well, that's just not true. There's as much water on the planet as there ever was. Where do they think this water went?
They talk about how people are dying of thirst.
Don't forget nonhuman people.

But these people ~ including nonhuman people ~ aren't dying because the world is running out of water. They're not even dying because they're running out of water.
They're dying...
Because the water is being stolen.

We're always told that the "solution" to the "water crisis" is that we should take shorter showers. But what we're not so often told is that more than 90 percent of the water used by humans...
Caw!
...is not actually used by humans at all, but by agriculture and industry. And of the water used by cities, as much water is used for golf courses as is used by human beings.

So the people aren't dying because there isn't enough water. They're dying because the water is being stolen, for mining and factories, for bottled water and soda pop, for cash-crop farms and ranches.
For golf courses.
So if you really want to make it so people have enough to drink...
Including nonhuman people...
...I think there are more effective things you can do than take shorter showers.

Like eat fewer French fries!
???

I read that salmon are dying in some places because the water is used instead for farming potatoes.
And in other places it's for factories, vineyards, orchards.
Caw!

So if we just eat fewer French fries, they will use less water and there will be more water for the fish, and the salmon will all be okay!
You eat enough French fries to make this difference all by yourself?

No silly! We all have to do it!
That will never happen.

We need to dream big, or nothing big will ever happen.
I saw that on television.
If we're going to dream big about everyone doing something, why dream so small as having them eat fewer French fries? Why not dream so big that everybody takes out the dams that are killing the rivers in the first place?
You're being silly again.
Hmm...this gives me ideas.
Caw!

The energy needed to manufacture 14 plastic grocery bags could power a car for one mile!

I need your help.
When have I not been there for you.
Never.
Never. What do you need?
Overconsumption...
There is no such thing.
...and our whole way of doing business and our whole way of life is killing the planet.
I don't like where you're going with this, Ed.
Hear me out.
No.
But you're killing me here. You're eating into my profits even talking about this stuff.
After all I've done for you. Everybody told me I should support your opponent in the last election because he's for removing every bit of regulation. But I took a chance on you. Don't fail me now.
I won't. I've got a win-win situation for us.
I'm listening.
People are worried about the environment.
Nobody I know.
True, but there are some... and people hate corporations.
What are you talking about?
Strange, I know, but trust me on this one. People hate corporations almost as much as they hate politicians.
But why? Oh, don't answer. I know why. They're envious of our success and they're too lazy to actually do any work themselves...
Oh, just a second, I'm going to buzz the secretary and have her make us some coffee and get us some pastries. Wouldn't that be nice?

So here's the plan: I'm known as something of an environmentalist...
I've never understood that.

I think you will now. As such, I'm positioned to help make you a leader in environmental sustainability.

You just said the system's killing the planet. Which means if we don't want to kill the planet we need to get rid of the system. But I like the system. I don't care if it is killing the planet. It's making me rich.
That's just it. I've got a way for "sustainability" to make you even richer.
People need to buy their low-energy light bulbs somewhere.
Why not from you?

Hmm, why not from me? But those bulbs are expensive.
Pass those expenses on to the consumer.
With me taking a cut?
With you taking a big cut. Greenies will pay outrageous prices just so they can say they're "doing the right thing."
I like how you think.

We can also get some subsidies for you to get solar panels or biofuels.
I don't pay extra?
Not a dime. You continue with business as usual.
And what do you know, I'll become the leader in sustainable overconsumption!

WOW! It looks so... so....
So delicious!

The first thing we'll have to do is set up a bureaucracy to facilitate resource extraction...

You're not going to believe this...
Don't tell me they're prepared to resist.
Quite the opposite. They already have bureaucracies setup.
What for?
For everything!
Including...
Yep!
To facilitate resource extraction?
But why?
I don't know. I just know it makes my diodes all tingly even to think about it. Can you believe they already set up the infrastructure for us to eat the planet?
Don't they know they have to live here? Everybody knows that you don't kill the planet you live on.
You only kill the planets you conquer.
Their loss is our gain.
This is all too easy. There has to be trick. I smell a rat.

I smell a human.
Of course. They're everywhere.
You can't get away from them.
And they're so disrespectful. Not like they used to be.
Disrespectful? More like murderous. Yes, they used to be fun, funny hairless creatures who played with us like we play with each other.
They used to sing to us.
Pray to us.
Now they're a bunch of psychopaths.
What are we going to do about that?

Aliens! Holy jumpin criminey!

Mr. President, aliens have landed, and they want to see you.
From Mexico?
ALPHA MALE
From outer space, sir.
Do they have documents?
I don't understand, sir.
Are they illegals?

I don't think immigration policies apply here, sir.
Because if any of them are illegals I certainly can't set up a meeting. It wouldn't be right.
Although Mrs. President could use another maid. We had to send the last one back to....

They're from outer space, sir. They've come to eat our planet.
Is that legal?

It is if you say it is, sir.
Well, why would I say it is?

It's better if they just tell you, sir.

They're not threatening me, are they?
No.
Because I won't be threatened. Especially by aliens. Aliens without proper documentation.

Shall I send them in, sir?
One more question.
Yes, sir?
Do you know if they've accepted Jesus Christ as their personal Savior?

Come again, sir?
Do you think Jesus died for their sins, too?
I... I haven't asked them that precise question, sir.

Probably not. They're aliens, for crying out loud. Undocumented. Probably illegal. I can't see why Jesus would die for their sins.

It's solar powered!
Groo-vay!

Oooh, fox-ay!

I've been thinking...
Not again!

I've noticed that you eat lots of berries and seeds and even grasses.

But I've also noticed that you eat poor little mousies.
You catch them. You grind them with your teeth. You swallow them. You digest them!!!

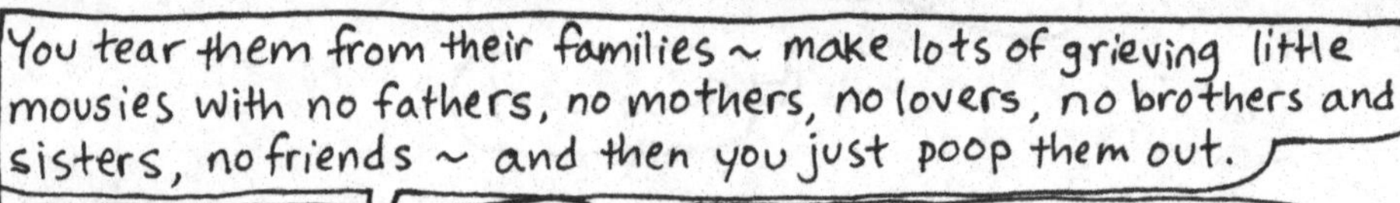
You tear them from their families ~ make lots of grieving little mousies with no fathers, no mothers, no lovers, no brothers and sisters, no friends ~ and then you just poop them out.

That's really disrespectful. And it's cruel. And it's not sustainable.
When a mousy eats a plant, only ten percent of the calories in the plant is converted into a mousy. The rest is wasted.

So it's ten times more efficient for you to eat plants than it is for you to eat mousies. And the planet could then support ten times as many foxes!

I think from now on you should eat only berries. Then the world will be a better place.
Only berries?
Grasses and seeds too.

You and I both know that mousies are sentient.
You think carrots aren't?

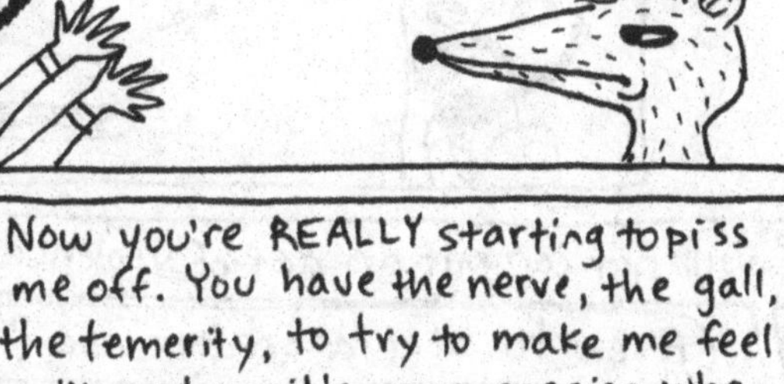
And I feel a personal responsibility to not add to the misery of other sentient beings. Don't you? You could set such a good example for everyone everywhere.
If you could go vegetarian~ and we know you could~ then anyone could. It's your responsibility, and if the tortures don't stop, it will be in part because you choose not to do your share.

Okay, now you're starting to piss me off.
Don't get angry. Just say calmly, Peace before me, Peace behind me.

Now you're REALLY starting to piss me off. You have the nerve, the gall, the temerity, to try to make me feel guilty when it's your species who created vivisection? You want to talk about torturing mousies? You want to talk about tearing them from their families? You want to talk about systematic torture?
No! No! I don't want to talk about that!

I didn't create factory farming.
I didn't create vivisection labs.
Nobody did but you humans.
Instead of getting us to make these little lifestyle changes, why don't you storm the vivisection labs and release the mousies?
Why don't you burn those places to the ground?

That would be violent!

Why do you hate violence that frees the victims of greater violence, even more than you hate the original violence?

I will not commit an act of violence.

Even to stop a greater one?
It seems like you don't want to be responsible for anything.
And by the way, I don't hate violence. I just dislike it very intensely. Hatred is destructive. Just like the violence you're suggesting is destructive.

You won't admit that carrots are sentient, but now you call taking down a building "violence"? Why? Are buildings alive?
It's violence against property!
And what's so wrong with that?
Don't say that. Don't say that.
I can tell you how to stop the torture of mousies...
I don't want to hear this.
Just go to the buildings...
I can't hear a word you say.
And release the mice...
Peace above me...
And then wait for the vivisectors to arrive...
Peace below me...
And if you have decent teeth or claws like most respectable animals you could just tear--
Nooooo!

I found another list! Do you remember that old book "Fifty Simple Things You Can Do to Save the Earth"?
Yes, things like if you somehow stop receiving junk mail, you could save one and a half trees per year?
It didn't work, did it?
365 Things

That's because there were only fifty things. The new book has 365!

Here's a good one: "Use concentrated dishwasher liquid."
Do we—

~Wait, here's another one. It's even better! "Decide what you want before you open the refrigerator door."

What?

You don't like this list either, do you?

If you think lifestyle changes won't make a difference, why are you a vegetarian? Why do you base your garden on principles of permaculture? Why do you bother having a composting toilet if there's no point?
360 Things

I never said there's no point to living responsibly. But I don't delude myself into believing that what I have for lunch, much less which dishwashing soap I use, can stop the system from destroying the planet.

You just don't like anything. Sometimes it feels like you're a negative person. Sometimes that negativity gets to me.
I don't like these lists. They're worse than useless...

But if we all did everything on the list...
It wouldn't be enough. And that's why the lists are harmful. They give people the illusion that the problems we face are easily solvable... Fifty simple things... The book should be called "Fifty Simple Things You Can Do to Stay in Denial While the World is Destroyed."
It's not that simple.

You're really negative.
It's not that simple.

It really is simple.
But the humans will never do it.
Yes, I know.
Not the domesticated ones.
Caw!

But it really is simple, isn't it? Why won't they do it?
You know the answer to that one.
Caw!

Did you notice that on the list from the end of that movie, all their suggestions for action have to do with individuals?
That's because they're things you or I could do.

But there was nothing about stopping the governments and corporations that are the main causes of the problems.
Did you know that all by itself, ExxonMobil has released five percent of all carbon emissions put out by this culture?
They're the real problems, not us.
But we can't do anything about them. We can only do something about ourselves.

That's what they want you to think.

Do you really believe they're that controlling and evil?
Yes.

You really need to do something about that suspicion and anger.
Of course I do. I need to stop those who are killing the planet.

No! No! No! You need to do something about the suspicion and anger, not with it. You need to see a therapist.
A therapist?
Yes. You just think the worst of politicians and CEOs. They're not as bad as you say.
They aren't?
You're so angry. Therapy would help you feel so much better.

I saw your film.
And?
I didn't much like it. Except the end.

Where I tell people what to do?
Brilliant... You should have been a stage magician: "Look over here, folks. Abracadabra."

And meanwhile...
The real action is somewhere else.

We're here to eat your planet.
Why would you do that?
Why do you do it?
BIG MAN
It makes money.
We think it tastes good. Judge for yourself which motivation is better.
I don't even have to think about that. Money is better.
We understand you have bureaucracies—
Cough cough!
er, we mean agencies, set up to facilitate the destruction of the planet
Cough cough cough!
er, we mean, set up to wisely manage this planet's resources to protect the public interest.
That's true, we do.
We understand how busy these agencies must be destroying—
Cough COUGH!
er, we mean managing these resources. And so we thought we'd come directly to you to expedite the permitting process.
You don't have permits?

That's why we came to you.
You're illegals?
? ?
And you probably don't believe in God...
You mean the Great Mechanical One?
Who so loved the world he sent his only son...
His scale model replica.
To die
For our sins.
To be shorted out
For our hard drive crashes and system failures.
We speak the same language, even if you are illegals.
About the permits so we can legally eat your planet...
Unfortunately that's not my job. I delegate.

Can you point us to your designated decision-makers, then?
We want to give them this.
Gold! Where did you get all this?
I pulled it out of my a
COUGH COUGH!
TOP BANANA
I suppose I could expedite the permitting process for you. It's nothing you need to worry your mechanical heads about.

You feel as though you're not well-adjusted?
I suppose that's a fair way to put it.

Then I'll just write a prescription...
Hold on.
Can I ask you something?

Certainly.

When someone is "well-adjusted," it means that they can tolerate whatever the norms of their society are, without being constantly upset and angry, right?
Yes, in essence, that's the therapeutic goal. For the patient to be at ease within the overall social structure and customs ~ not only to be free from pain, but even to be...um... happy.

And these pills you plan to prescribe will help me to be well-adjusted ?

They can help, yes. They alter the chemistry of your brain, to smooth out the ups and downs of everyday life, enhancing enjoyment and productivity.

Productivity. Hmm.

So it will no longer bother me that our planet is being destroyed by greedy corporations?

I won't cry any more about the frog species going extinct, or the polar bears swimming through the ocean to ice that's no longer there?

I won't be haunted any more by the ghosts of the numberless war dead, soldiers and civilians, lost friends and broken families?

The agony that I feel about all the people being kidnapped and legally tortured by the government will ease off ?

The screaming won't trouble mydreams any more ?

The rage I feel about people being worked to death and still being unable to feed their families or afford health care, this anger will subside?
All these bad feelings will magically vanish with a little pill, and I'll be happy and adjusted to this gruesome reality, this ravaged planet and suffering people?
Is that what you're saying?

My sorrow, my empathy, my caring, my sensitivity ~ you want to kill all that so I'll be PRODUCTIVE, so I'll be more useful to the big capitalist machine, do I have it right?
Why not just cut out my heart and be done with it?

One moment.
Symptoms:
depression, paranoia, victimization/helplessness, anger issues, delusions of grandeur.
Diagnosis:
Totally fucked up.

I'm going to give you a bit of a higher dosage than I planned initially.

And if I don't want pills?
We, uh, we can work on your issues.

You seem very suspicious. Why did you come in the first place?
My friend Bananabelle asked me to.
I see.
Displacement of responsibility
She thinks I'm too angry.
Are you?
No.
I can't help you unless you admit there's a problem.
There is a problem. It's just not with me. I have to find a way to stop the horrors.
What horrors? I don't see any horrors. I see a nice young woman sitting on a nice chair in a nice room. What could be nicer?
A world without vivisection.

I don't see any vivisection.
If it's not in the room, it doesn't exist?
I'm trying to get you to be in the here and now.
And the here and now includes our social circumstances, which in this case includes the fact that this culture is killing the planet.

You try so very hard to be unhappy. That must be exhausting. I'll bet you're tired and you just want to rest. I'm here for you to help you relax, to help you enjoy this moment.
Did you know that there are more than 200 dead zones in the oceans caused by the actions of the industrial economy?
Let's stay in this room. Do you see any dead zones in this room?

Seabird populations are plummeting across the planet.
I don't see any albatrosses here. I'm wondering what it is about you that causes you to not want to remain in this room. Are you uncomfortable in your body?
My body is fine. My body doesn't make me angry. Vivisectors make me angry. Factory farms make me angry. This culture makes me angry.

No one can make you angry without your permission.
If I punch you, that won't make you angry?
Do you want to punch me?

You need to breathe out those bad emotions, and breathe in only good emotions.
Anger against injustice is a bad emotion?

You're insane. Bonkers. Stark raving mad. You're crazy, and you're trying to drive me crazy.
I can't drive you crazy. No one but you can do that. Why would I want to drive you crazy?

If I'm a sane human being, I'll remind you that once you were a sane human being too, that you were an animal who felt emotions and who felt outrage at a system that is killing everything you hold dear.

It's not killing everything I hold dear. I'm quite happy. You're the one who seems upset.
You're too afraid and too small to even feel outrage at the system turning you into a drone.
Interesting. I don't see any "system." I only see you and me sitting in this room, talking about your anger. I see and hear you acting out a lot of hostility. Are you always like this?
Are you sure your anger really is at vivisectors? Are you sure it's at this thing you call "the system"?

How is your relationship with your father?
My father's got nothing to do with it.
Why don't you want to talk about your father?
Whether my father was wonderful or flawed or horrible is entirely independent of the fact that right now vivisectors are torturing monkeys, rats, dogs, cats, mice, birds, insects, that right now sea turtles are being driven extinct, that right now amphibian populations are collapsing...

How does your anger affect your social life?
I don't want to talk about my social life.
Are you uncomfortable talking about your social life because your anger drives away your friends?
I'm not angry at my friends. I'm angry at the things that make me angry.

How do you know that "the system" is not just an easy place for you to put the anger you really feel toward other people, and toward your difficulties with other people? How do you really feel about your father?
If my father were perfect the system would still be killing the planet.
Ah, so you acknowledge your father isn't perfect.

I'm sure a smart girl like you knows what denial is, right?

Denial can run very deep.
Yes, it can.

And it can cause a lot of pain for those unfortunate enough to come in contact with people in denial.
I am perfectly aware of what you are saying.

Well, do you want to stop causing the pain?
I want to stop the culture from Killing the Planet.

You seem to be obsessed with destructive fantasies, with what you call "dead zones" with what you call "torture." That's a lot of negativity to carry around with you. That would be awfully heavy. I would want to lighten that load by setting it aside.
Yes, you would.

And I'm concerned about all this affection you have for sea turtles and such. Sea turtles aren't your family. What do sea turtles represent to you in your own life, in your real life?
Sea turtles are real life. They aren't just projections of my issues.

It can be scary to care about another human who can reject you. Sometimes people are too scared to even acknowledge this fear, so they project this fear of personal rejection by other humans onto things like "sea turtles" and their supposed destruction by some "all powerful system."
At some point you have to give up those fantasies and live in the real world.
Sea turtles aren't the real world?

You see this big scary world, and you think it's going to destroy you, and you want someone to take care of you just like you say you want to take care of "sea turtles." But you'll never find someone to take care of you as long as you are this angry.

I don't want someone to take care of me.

I think you do. I think we all want someone to take care of us. And that's why we created this whole system. It gives us food and shelter and air conditioning in the summer and central heating in the winter.
No.
Just let go of your anger. Don't you see how pointless all of your anger at this "big bad system" is?

Why?
Because ultimately you can't do anything about it. So just lie back and enjoy what you can.
You'll be ever so much happier.

I get it now. I understand what you are doing.
I'm just trying to help you.

You're trying to get me to believe I'm helpless, that people really can't change things, that people really can't stop this culture from killing sea turtles and everything else.
Because if you can convince me not to act, then maybe your own inaction won't seem as horrid as it really is. And if you can convince me not to act, then I might not bring down this system that brings you tomatoes in January and air conditioning in July.
But more important than that, if you can convince me I'm helpless, maybe you can convince me I need your help.
How convenient for you!
Your fifty minutes are up. You can pay the receptionist on the way out.
Would you like to schedule another appointment?
Clearly paranoid, and a danger to self and others.
She needs to be apprehended, for her own good and for the good of society

Excuse me, sir.
Not now, I'm busy.
I know that, sir. But...

This better be an emergency.
The aliens are consuming the Great Lakes.
I thought you said this was an emergency.

Well, some seem to think so, sir.
Who?
Some of the people who live there, sir.
Why haven't they been arrested?

The aliens?
No! The people! The complainers! The whiners! Those who oppose our programs! The terrorists! If those people don't agree with our policies, they are terrorists who must be arrested, and I mean now!

But they're consuming the Great Lakes, sir.
The people?

No, the aliens, sir.
I thought you said this was an emergency.

Sir?
Take a look at the aliens' permits. You'll see that they're in order.

Permits in order, sir?
Signed by me personally. Now go away. I was just about to take over the world.

BRUTE

Kill the bad guys!

Yes! Kill! Kill!

Excuse me, sir.
Not now. I'm reading my Bible.
But sir...

What?!?!?
The aliens are eating the Appalachian Mountains, sir.
And?

They're chewing off entire mountaintops, sir. I just thought you'd want to know.
And have you arrested the people who complain?

Umm, no, sir.
what do I pay you for? Now get busy.
I need to go back to reading about Jesus and his benevolent love.

Sigh.

Umm, sir? I'm terribly sorry to bother you.
It's the aliens again, sir.

They are eating every fish in the ocean, and every tree in the forests.
So what's your point?

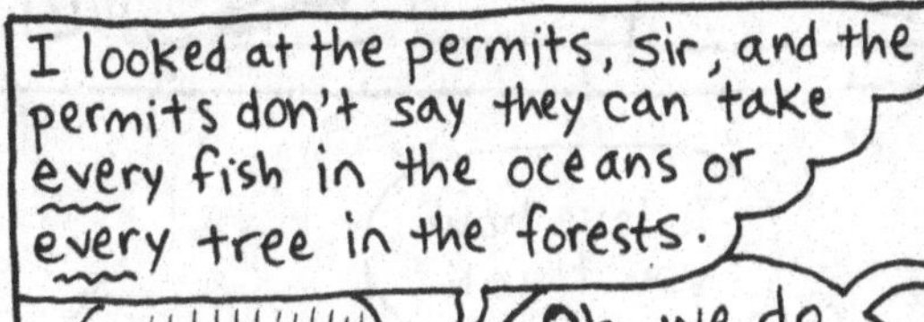
I looked at the permits, sir, and the permits don't say they can take every fish in the oceans or every tree in the forests.

Oh, we do have a problem. Who pointed that out to you?

Some people who lived near the forests.
Arrest them immediately!
About the permits, sir.

What do they say?
They say some fish, sir, and some trees.
I don't think we need to worry about a few fish or trees.
They're taking them all, sir.

Some, all. Whatever. As long as they've got the permits. Signed by me. If I sign them that makes them legal. And if I say they can have all the fish, they can have all the fish. Is that clear?
Extremely clear, sir.
I'm glad that's settled. Make sure to arrest anyone who disagrees.

This is all too easy.
I almost wish they would resist.
Not too much.
Just enough to make it a challenge.
But not enough to actually stop us!
I love how you think.
Ooh, you're making my diodes tingle again!

The store doesn't give us food.
The Earth gives us food.

We've got to stop them!
They're killing the planet.
Oh, don't go overboard. It hurts the cause when you exaggerate like that.
They're killing every fish in the ocean and every tree in the forest.
They're not killing the planet. They're not strong enough to kill the whole planet. Algae or something will survive.
That's right. Besides, doom and gloom doesn't sell. It turns people off. We've got to make sure we give people a message of hope, give them something they can do.
Like blow up the aliens?
You know better than that. You can't just blow them up. Gandhi would never have blown them up.
Besides, they have permits, signed by the President himself.
So how do we stop them?
ARCTIC ECO-TOU

As you know, the permits only allow them to take some of the fish in the oceans. They're taking them all.
I think we can stop them on procedural grounds there.
What are we going to do?

Well, we've got our petitions we can start circulating, and I wrote a letter that we're going to mass-mail that people can sign and send to the President.
Did you write a fundraising letter we can send at the same time?
That's exactly the sort of thinking I like to see.

Let's go get 'em!

I don't know how much more of this I can take.
You and me both, brother.
I mean, what's the use of a legal system that allows destructive behavior?
Ha! That's the point of a legal system: to legalize destructive behavior.
By the rich, that is.
What are you talking about?
Who makes the laws?
The rich and powerful.
Who benefits from these laws?
The rich and powerful.
What is there not to understand?
When they destroy the planet not even the rich and powerful benefit.
They just think they do.

What will happen if we don't stop them?
They will kill everything there is.
Why don't we stop them?
I don't know. Why don't you stop them? We're already trying as hard as we can. You're the ones who aren't pulling your weight.
You're really smart. What would I do without you?
You would die.

How's the campaign going against the aliens eating every tree in the forest and every fish in the ocean?
Great! Over ten million people have sent our letter to the President!
And we've raised over five million dollars for our organization!
And we've got our petitions circulating.
What about the trees?
What about them?
We even have the names of some massive stars. MASSIVE!
What about the fish?
Do you think it would be more effective to re-arrange the signatures so all the celebrities are on top?
The campaign is going great!

What about us?
What about us?

It's nice and toasty in here.
Yes, we have a new source of renewable energy.
FIRST LADY
FOOT BINDING
FASHION
TORTURE FOR DUMMIES
What's that?
Letters and petitions.
How nice that you keep them out of landfills.
It's good to be green.
Do you read them first?
Are you kidding? No, right into the furnace they go.
You don't even take down the names to arrest them?
I thought about that, but I want to encourage this sort of behavior...
So people will think they're making a difference?
I hadn't thought of that. I was just thinking so we'd have enough fuel to keep us warm.
FIRST LADY
Fascism for Dummies

And why do you have that pile over there in the bathroom?
I use them for...

You DO? Aren't they rough and unabsorbent?
Yes, but I do like the symbolism.

That's my Mr. President, always willing to suffer to express his convictions.
WAR FOR

You seem quiet today.
I'm afraid to say what I'm thinking. If I do, you'll just tell me I'm wrong.
Oh, I'm sorry. I never meant you were wrong. All those suggestions you've made have been really good! It's just that they're not enough. And they put the focus on us, instead of where the real problem is. And they keep us from doing what we all know needs to be done. And they...
You're doing it again.

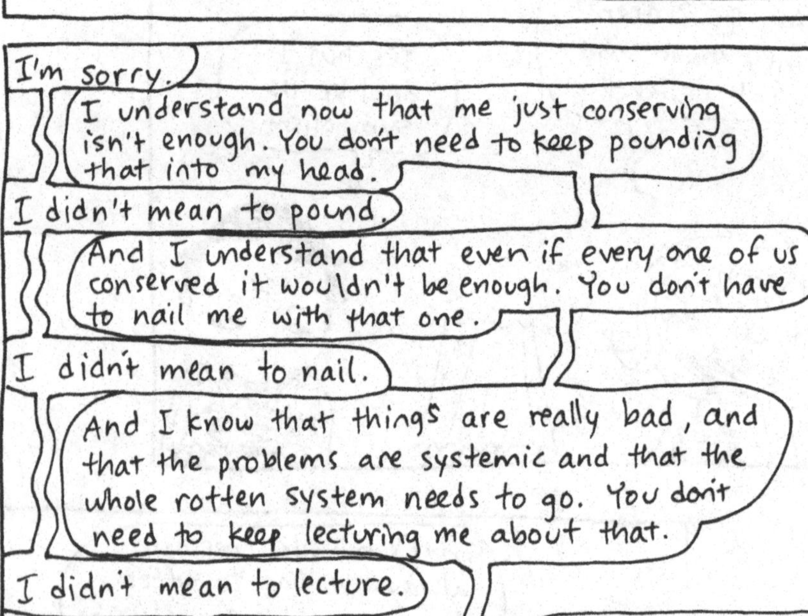
I'm sorry.
I understand now that me just conserving isn't enough. You don't need to keep pounding that into my head.
I didn't mean to pound.
And I understand that even if every one of us conserved it wouldn't be enough. You don't have to nail me with that one.
I didn't mean to nail.
And I know that things are really bad, and that the problems are systemic and that the whole rotten system needs to go. You don't need to keep lecturing me about that.
I didn't mean to lecture.
But can't we keep things like air conditioning and cars and radios and TVs and computers and just get rid of the bad stuff like pollution and exploitation?
Wouldn't solar panels or windmills make it so we can have that stuff?

Oh, I know what you're going to say. You're going to say that in order to have air conditioning and all those other things ~ even if they're solar powered ~ you still have to have mining for the copper wires and the silicon and everything else, and you have to have the whole industrial infrastructure, which means you have to have roads and the whole oil economy to move stuff around, which means you have to have this huge military so you can *take* the oil from wherever it is, and you have to have this whole system which leads to some people being rich and some people being poor, which means you have to have police to make sure the poor don't take back from the rich, and you have to have prisons and everything else that comes with that.

And then the rich keep getting richer and the poor keep getting poorer.
And it all keeps killing the planet.
I know that's what you'd say.

So then I would say, what about high technology? Can't that work? I really want it to work. I've heard all these great things about how nanotechnology will make production so much easier.
Wouldn't that be a good thing?

But then I know just what you'd say. You'd say that all these technologies follow the same pattern of being hyped and hyped and hyped, and then they always ~always ~ cause more pollution and they always ~always ~ cause the rich to get richer and the poor to be more exploited, and they always ~ always ~ cause the world to be hurt more and more. Every time.

I know that's what you'd say.

And then after that you'd ask: who controls these technologies? Can I make nanotechnology in my basement? And then you'd say, Of course not. Big corporations will control it just like they control everything else.

And then I'd get really sad because it won't work the way it's promised and I want it to work, and I'll know you're right but I don't want you to be right.

So then I'll ask if we can use biofuels to save us. And you'll say that industrial agriculture is based on natural gas and oil for fertilizers and pesticides, and it's incredibly toxic and it destroys soil and water.

And you'll say it's all controlled by big corporations.

And you'll say that if every single bit of cropland in the United States was used for biofuels it would make up only fifteen percent of this country's demand, but even then it's impossible because growing corn for biofuels actually takes more energy than it makes.

You'd tell me that currently it takes ten calories of dino-fuel to grow one calorie of food in the United States. People have covered croplands with McMansions because they believe they currently have that fuel to waste.

But then you'd tell me that without the ten calories of fuel, they will need way more laborers and way more croplands to feed the people who currently exist.

And then I'd be really sad again because every part of me wants for it all to be really easy, but you'd be right again, and it wouldn't matter that you are right but what's important is that I would understand that the problem really is the whole system but I wouldn't want to believe that and I'd rather believe that all our problems can be solved if we just buy energy-saving lightbulbs and if we just recycle.

See, I told you that if I said those things that you'd shoot down all my suggestions. I told you that you'd just make me feel bad.

I didn't mean to.
I know.

Am I still your friend?
Of course. My best friend... even if you do tell me those things.

Mr. President, you have a visitor.
Another alien, with gold?
Umm, no sir. It's the President of Mexico, sir.

Mexico? Does he have papers?
Yes sir, he's the President.

Ah, so I suppose he can give himself papers.
You're propably right, sir.

Did he bring any gold?
I don't believe so, sir.

Then why should I see him? I've got lobbyists lining up outside the door waiting~ waiting~ to pay good money to see me. Why, just the other day Mrs. President suggested I start auctioning access to me on Ebay. Like my daddy used to say, time is money, my boy. Why should I give it away, even to the President of Mexico?

He's concerned about the aliens, sir.
But why?
They're eating all the forests.
They've got permits.
The permits are only good for this country, sir.
The permits are good wherever I say they are. I'm the President of the United States! This country extends all over the world, doesn't it?
Umm, not exactly, sir.
Well why not? Whose fault is that?
I don't know, sir. You might have to ask the other heads of state about that one.
I would, if only they would come see me and bring me gold.
I'll contact them about that, sir.
In the meantime, tell the President of Mexico to make his own deal with the aliens. I'm sure they'd make a deal with him, too.
Like my daddy used to say, why sell a permit once when you can sell it again and again?
Very good sir.
Of course it's very good. It's from my daddy, who knew a thing or two about selling permits.
Well, why are you just standing there?

The President is very concerned about the fish in the oceans.
No, I'm not.
The President of Mexico, sir.
Why would he be concerned about fish in the oceans?
They're disappearing, being eaten by the aliens.
So what? That's what they're there for, isn't it? And the aliens have permits, bought and paid for.
He said to tell you that he personally doesn't care if the fish are all gone.
Good man.
But poor Mexican fishermen are being driven out of business.
I still don't see the problem. That's just part of global economics. Big corporate fish eat little inefficient mom and pop fish. And in this case the aliens eat all the fish.
That's the way the System is supposed to work.

But they're very poor, sir.
And why exactly am I supposed to care about that? They obviously weren't very good fishermen, were they?
Not like my daddy. You should have seen him fish. He could catch him some bass like you wouldn't believe. And big ocean fish, too, when we'd rent these nice charters where they bring you dinner and drinks, and bait your line for you and even reel them in once they're hooked.
Now he was a fisherman.
The Mexican fishermen are starving.
Tell me again why I'm supposed to care? These fishermen probably didn't even have permits anyway. They should be glad we didn't arrest them.
They're in Mexico, sir, you couldn't have arrested them.
Who would have stopped me?

They're marching on Mexico City, sir.
And they haven't been arrested?
No, sir.
Oh, you know, that's very wise on Mr. Mexican President's part. This is very good. All these former fishermen will create a greater labor pool and bring down wages even further in WalMart's Mexico City stores.
Get on the phone to my stockbroker and tell him to buy a million shares of WalMart stock, and I mean now!
They don't want jobs at WalMart, sir. They want their old livelihoods back.
They probably want to turn back the clock and make it so the Red Sox won the World Series, too.
The Red Sox did win the World Series, sir.
Why doesn't the President just sell permits to the aliens for the fish too?
Because the fishermen won't be satisfied with that, sir.
I don't understand why the President of Mexico cares whether some poor fishermen are satisfied. That's not his job. This isn't some popularity contest.
His job is to sell permits, and if he doesn't understand that, then he needs to get another job, and we need to get another President of Mexico.

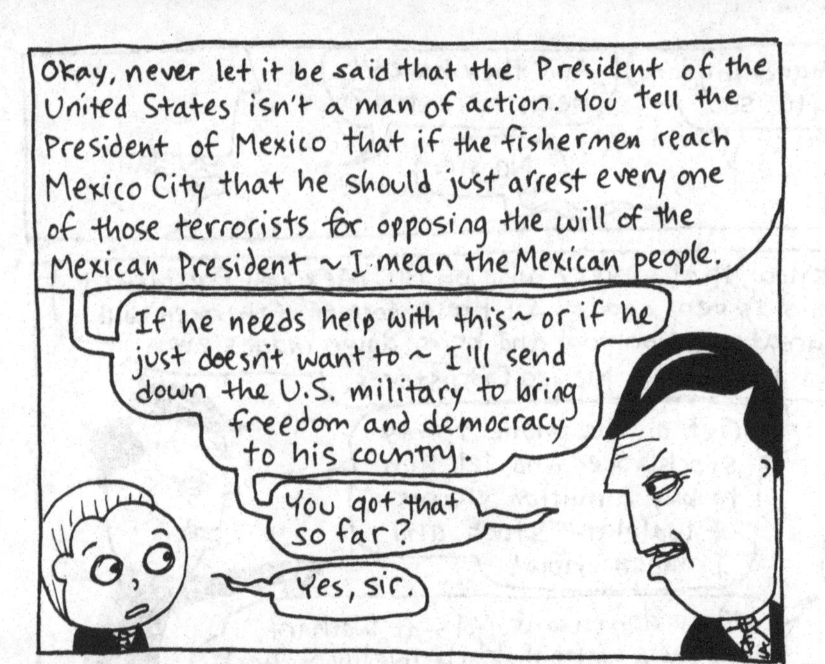
Okay, never let it be said that the President of the United States isn't a man of action. You tell the President of Mexico that if the fishermen reach Mexico City that he should just arrest every one of those terrorists for opposing the will of the Mexican President ~ I mean the Mexican people.
If he needs help with this ~ or if he just doesn't want to ~ I'll send down the U.S. military to bring freedom and democracy to his country.
You got that so far?
Yes, sir.

And then you tell him to tell all the other little countries down there ~ I never can remember all their names ~ the same thing. If they have any trouble at all with the people, we have troops available specifically to bring freedom and democracy to their countries.
You tell him that.
Yes, sir.

And one more thing. Tell him also that when he gets back to Mexico City, to send up one of his aunts.
Mrs. President still needs another maid.

I'm concerned.
Concerned?
About the trees.
Trees?
And the fishes.
Fishes?
They're disappearing.
You're concerned about trees and fishes disappearing?
Will you quit repeating everything I say? I understand that's your job as a politician, but right now it's not helpful.
I just didn't know you cared about trees or fish.
You think I care about trees or fish?
But you just said...
I'm concerned because the aliens are getting them.
But I thought...

Ed, WE are supposed to get all the trees and fish. How are we supposed to make money if the aliens get them all first?
What about bushes? Can't you get those?
The aliens are taking those too.
Rocks?
Everything.

Don't they know corporations are supposed to get everything?
What's wrong with them?

Hey, I've got an idea. Can't you just buy things from the aliens and re-sell them?
Use your head, Ed. Buy them with what? They've got so much gold it's almost as though they're pulling it out of their~
So what do we do?

When buying doesn't work, what is the next step?
We stop them.
Exactly. We stop them.

We've got to stop them.
How?
The letter-writing campaign to the President is almost there. I can feel how close we are to convincing him.
We've raised a lot of money, too.

I think we need to do something to push our effort over the top.
We need to raise the stakes.
Exactly. And the way to do that is through press releases.
And protests.
So long as they're orderly.

And we need celebrities!
Oh, yes. Singers. Actors.
Beautiful actresses.
With great ti
Is that all?

No, of course they've also got to have sweet as
No, is that all we're going to do?
All? Are you kidding? Letters, protests...
Don't forget the petitions.
And actresses.
What more could there possibly be?

We've got to stop them.
But how? They're a bunch of machines.
We just do. That's all.
We use whatever we've got.
Fight like our lives depend on it.
They do.
Will the humans help?
The humans are as bad as the machines.
The humans have made themselves into machines.
Not all of us have...
Not all of you.
But the machine-humans: they won't help.
No, they won't. They hate life.
Some of them might help.
Don't count on it.

Let's go.
Use whatever you've got.
Use everything you've got.
Like your lives depend on it.
They do.

How's the precious little babybot?
The little snookems...
The chip off the old mainframe...
Growing new gears every single day.
How many snookems is that now?
How could you forget something so important? If you don't remember I'm certainly not going to tell you.
In other words, you've lost track too.
Yes, isn't it wonderful?... Do the calculations again, will you? It makes my diodes tingle when you do them.
Well, let's see, if you count our own snookem babybots, and our snookem babybots' snookem babybots, and their Snookem babybots, and on down the line, and if we reproduce at a rate of... hmmm, well, I've got us at six million, four hundred and forty seven thousand, nine hundred and thirty four, oops, thirty five, oops, thirty six.
The Great Mechanical One must be so pleased.
How many now, darling?
6,447,943, oops, forty four, oops, forty five...

Mr. President, we have a problem.
You bet we do!
You agree?
Several problems, in fact. One is that I can't get past level seven on the new game I'm playing. But you didn't come about that, did you?
No, sir.
Another is that I can never get enough gold. But you didn't come about that either, did you?
No, that's a problem we all have.
Ah, I know what problem you came about. A big one. Life-threatening.
That's the one. It has to do with aliens...
Oh, we've got that one solved. Mrs. President found this wonderful maid, an El Guatemalan or Nicarican or something.

I meant the other aliens.
The ones overruning the planet?
Yes!
the ones threatening our way of life?
Yes!
The ones who aren't really human like we are?
Yes!
The ones without permits?
Ye—umm, no.

May I be blunt, Mr. President?
Certainly. My daddy taught me to always prefer dealing with people who are really blunt instead of people who are really sharp. You're not sharp, are you?
Not particularly.
Good.

You know that the aliens ~ the machines ~ are consuming everything on the planet, converting everything into more machines...
Yes, of course. I'm the one who signed their permits.
You know they are destroying the planet...
Of course. Even people who are really blunt can figure out that converting a planet to machines destroys it.

And you know that machines are making it so nothing can live here...
Come on, Ed. I'm not stupid. Of course. Everyone knows that if you destroy a planet you can't live on it.

You really understand that they are killing the planet?
Yes! And?
And you don't see the problem?
Not at all.
You don't?

Ha! You're right, Ed. You really are blunt. Don't you get it? By the time they kill everything I'll be dead. Deceased. Terminated. Gone.

What do I care if the planet is dead after I'm gone?
That's it?
Well, I'm going to heaven, too, but in either case I'd rather have gold now than a living planet after I'm dead. Wouldn't you?
May I be extremely blunt, Mr. President?
Anything but sharp.

The problem isn't what's going to happen after you're dead.
That's what I said. It's no problem.

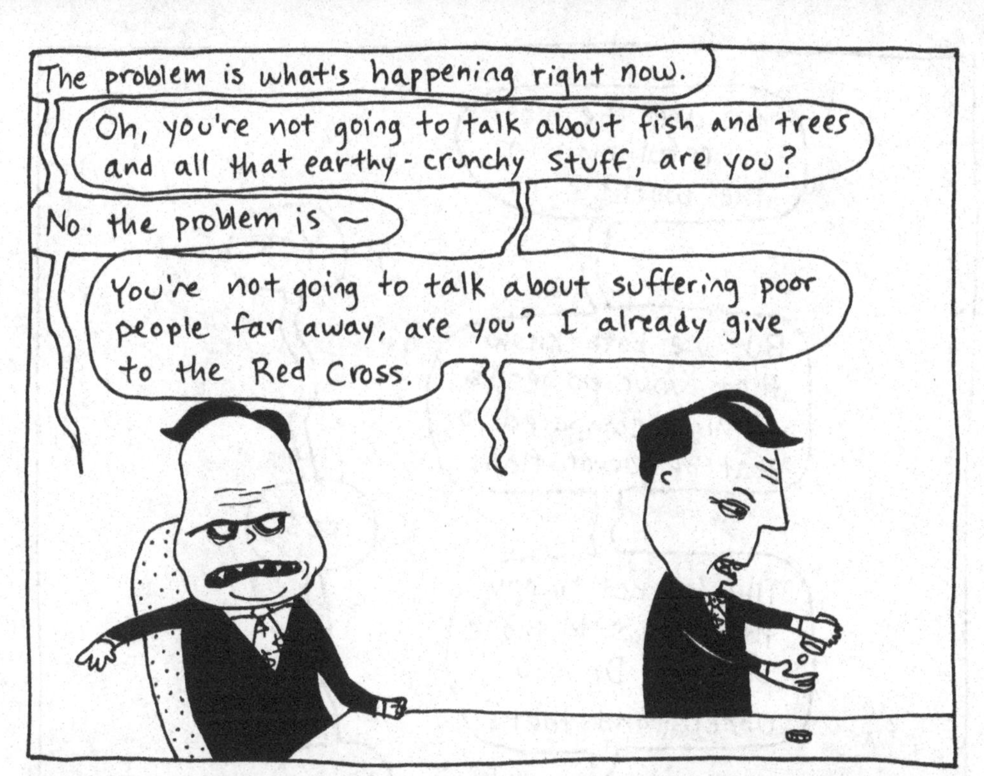
The problem is what's happening right now.
Oh, you're not going to talk about fish and trees and all that earthy-crunchy stuff, are you?
No. the problem is ~
You're not going to talk about suffering poor people far away, are you? I already give to the Red Cross.

No, the problem is that aliens are eating the planet.
Aliens aren't supposed to consume the planet, Mr. President.
Corporations are.

Good Lord, Ed. You're right.
But I already sold the permits. What are we going to do?

You are the most powerful man in the world.
Yes.
But we both know that your power is nothing compared to that of corporations.
Yes.
They're not happy that you sold those permits. Do you understand that?
Yes.
They decide who lives and who dies. You know that too, right?

Are they threatening me?
No more than they would threaten anyone else who got in the way of their profits. It's nothing personal, Mr. President.
They'd do the same to me if I stood in their way. Why do you think I serve them?

I understand how they work. I just didn't think it applied to me.
I've been told explicitly that it applies to everyone. Nothing~ and they mean nothing~ can be allowed to stand in the way of their profits.

Oh.

I would like to spend a little of this gold.

Do you love God?

Of course.
Do you love him so much that you would like to meet him tomorrow?

Now that you mention it, I don't love him quite that much. I, er, I prefer to love him at a distance.

You've got to get those permits back. The corporations need to consume the planet.

What are we going to do?
We've got to feed the corporations.
We've got to stop the aliens.

We have a report of an angry young woman. Refuses to listen to reason. Refuses to be sedated. Has been known to speak out against the social and economic system. Says it's killing the world. She is to be considered dangerous, and a potential terrorist. Be on the lookout. She has dark hair, dark eyes, and....
I'll get right on it.

Excuse me, miss.
Yes?
We have reports that you've been expressing hostility toward "the system."
You don't?
I, uh, well, what?
You have more in common with me than you do with CEOs and politicians.
What's that got to do with anything? My job is to keep the streets safe.
Then why are you coming after me?
I, uh, well, the dispatcher told me to. It's my job.

Your job is to keep the streets safe for CEOs and politicians?
The dispatcher said you were dangerous.
Do I look dangerous?
You might be a terrorist.
Nazis called the Jews terrorists. If your boss told you to arrest me because I was Jewish, or Black, or left-handed, would you?
My boss wouldn't ask me to do that.
But if he did...
I, uh, well...
If your job depended on it.
I've got a family to support.
In other words...

I, um, only arrest people who are a threat to society.
How many polluters have you arrested?
What? Why?
Has anyone you love been killed by cancer?
My grandfather and my cousin and my older sister. But what does that have to do with arresting polluters?
Cancer is mainly caused by the toxification of the total environment.
So how many polluters have you arrested?
Me, personally? Umm, since when?
Since forever.
Well, I, um, I, you see, I don't, umm, not very many. Lately, that is.
As in?
Well, none. Ever.
And you were going to arrest me because I express hostility toward the system...
I guess this time I don't have to arrest you. This time I can let you go with a warning.
Just don't do it again.

I've got ideas.
I've got ideas.

I'm sorry to bother you again, miss.
Yes?
I just got a report that someone blew up a dam.
They did?! How wond-

I mean, they did? Why would someone do something like that?
They left a note. Said they were doing it for the fish and the river.

I wonder who it was.
Fortunately there was a witness. Have you ever seen a one-eyed rabbit?
A one-eyed rabbit?
Noooooo, I don't think so.

And you had nothing to do with this?
Not this time.
What did you say?
I mean, no.

If you see a one-eyed rabbit, let me know.
If you see a CEO, arrest him.
What?
Never mind.

Bunnista, you're my hero!

Why? What did Bunnista do?

Bunnista was, um, playing with some, um, bunnies in a field and some people were treating the bunnies bad, and Bunnista stopped them!
You did? You stopped some meanies? You're my hero, too! Did you do it by telling them they were meanies? I'll bet that worked, didn't it? Didn't it?

Protecting bunnies from meanies. That gives me another idea!

Excuse me, sir.
Don't bother me now. I'm solving global warming.
This is important, sir.
More important than solving global warming?
Umm, no, sir.
Ah, there, I've done it. Everybody thought it was so tough. But I solved it all by myself.
How, sir?
Simple. I doubled speed limits all across the country.

I don't understand, sir.
Forget all that messing around with the carburetion fuel injectors or whatever they do to increase gas mileage. The answer is so simple that only I was qualified to see it.
I'm sure that's true, sir.
Here's the solution to global warming. If everyone drives twice as fast, their engines will only be running for half as long, thereby cutting all emissions in half.

Well?
You were right, sir. You are uniquely qualified to see that.
Now that we've solved global warming, what's so important?

Somebody blew up a dam, sir.
Why would somebody go and do something like that?
He left a note, sir, saying he did it for the fish.
For the fish?
For the fish, sir.

Damndest thing I ever did hear. Did he catch any? Maybe I'll have to try that sometime. Thank you for telling me that. You were right, that really is important.
He did it to save the fish, sir, not to catch them.

He blew up a dam to save fish?
Yes, sir.
Why would anybody do something like that?
Evidently he cared about the fish, sir, and the river.
Why would somebody care about fish or a river? He must be insane.

Did he have a permit?
No, sir.
He took out a dam?
Yes, sir.
For the fish?
Yes, sir.
Owned by a corporation?
Yes, sir.
Then arrest him!
We haven't caught him, sir.
Well then, arrest everybody.
Everybody, sir?
Well, not me. Arrest everybody it could have been.

A witness said it was a one-eyed rabbit.
That doesn't matter. He may have had accomplices. Cast a much wider net.
Meaning...
It's a fishing term. Since we're talking about a dam, I thought I'd...
I understand, sir. In this case, what does it mean?
It means arrest everybody.

We've been through this, sir...
You know who I mean. Everybody I don't like... Everybody I might not like... Everybody who looks like somebody I might not like... Everybody who doesn't like my plan to solve global warming...
On what grounds, sir?
Murder! Of course!
No one was killed, sir.
A dam was murdered! That's for sure. And they didn't even catch any fish.
I don't think we can charge someone with murdering a dam, sir.

Then terrorism! Get them for terrorism!
No one was terrorised, sir.
Surely someone was frightened.
Not that I know of, sir.
Not even scared?
No, sir.

How about startled?
No, sir.
Surprised, then?
I suppose so, sir.

Well, there you have it. Besides, it was a corporation's dam. Losing their dam will cut into their profits. Cutting into corporate profits is terrorism, pure and simple.

It is, sir?

Falling corporate profits sure scare me, and they scare everybody I know, everybody who matters. Let's round 'em up for terrorism.
I'll get right on that, sir.

If this planet dies, we'll just build space colonies. No problem!

I've spoken to the President, and we need your help.
Oh?
I knew our letter campaign was working!
It's huge. Bringing in lots of contributions.
Good...
And sending millions of letters to the President.
Yes, it's been quite warm in the President's office.
So we're really putting the heat on him, are we?
You might say that.
And now he's begging for our help?
Let's just say he recognizes that giving the aliens permits to take all the fish and all the trees and everything else was not in his ~ I mean the American people's ~ best interest.
The voice of the people was heard! Democracy works!
He wants to find a solution where everyone wins.
So do we!

So the President would like for you to mobilize a new letter writing campaign asking that he transfer the permits to take all the fish, trees, bushes, rocks, water, air, and so on away from the aliens...
All the permits? That's more than we dared dream. Yay!
And that instead he give those permits to the corporations who really should have had them in the first place.
Wait! What? Why?

You know you've got to give something to get something.
But what do we get out of it?
The President is willing to leave twenty trees, twenty fish, twenty bushes, and so on, out of the permits.
Twenty?

Those trees, fish, bushes and so on will be allowed to live undisturbed forever...
Twenty?
Or at least until the corporations need them.
I just don't know.

I'll tell you something in confidence.
Yes?
The President needs your help on this, so I think that number is negotiable. We can push the President hard on this.
How hard?
Very hard.

Do you think we can go as high as fifty?
He might go higher. If we shoot for one hundred, we might get seventy-five.
Just to be sure, that's seventy-five fish and seventy-five trees, right?
And bushes, and rocks, and so on.

The President won't jerk us around on this one.
You have my word.
You're a good man, Ed.
You drive a hard bargain.

In response to the recent terrorist act of blowing up a dam, today the FBI and local police around the country arrested 700 rabbits and rabbit supporters as suspected terrorists.

An anonymous eyewitness provided this description of the terrorist mastermind: "I saw a bunny slinking around, not hopping like most of them do. So I looked real close and saw he had one eye gone. Then I saw him plant the charges and tamp them down with his back legs."

When asked why he did not stop this terrorist rabbit from carrying out this terrorist act, the anonymous witness replied: "I did what anybody would do if they saw a one-eyed rabbit tamping down explosives next to a big dam. I ran like hell."

Our source at the FBI tells us that none of the arrested bunnies is one-eyed, but the FBI arrested them anyway for being members of the same species. Our source states, "A rabbit is a rabbit is a rabbit."

Government officials have raised the terror alert level to carrot orange, meaning that passengers will no longer be allowed to bring rabbits onto airplanes or other forms of public transportation. If you see a one-eyed rabbit we urge you to call the FBI immediately, or face imprisonment as a suspected terrorist and rabbit supporter yourself. No one knows where this rabbit terrorist will strike next.

We'll see who
are the real
terrorists.

Shhh, shhh. It's over now. You're going to be okay.
Oh, how could anyone...?
Hold on, I'll get you out.
I'll get you all out.
There must be a screwdriver or hammer or saw or something.
Something other than scalpels and knives.
I hope this will work.

I've got you.

Rest here. You're safe now.
I need to get the others.

Come.

Can you all make it outside? I'm going to check all the other rooms.

Oh no.
No no no.
Come out. Don't be afraid.

Can you all make it out? Help eachother; don't lose anyone.
Out to the left.
Everyone go outside.
Don't leave anybody behind.

Are you okay?
I have to check each floor.

Oh Kitty.
Go down the stairs and straight outside.

What's going on?

I always knew you'd come back for us.
I never forgot you for a single second.

This way, honey.
There are cages back there.
I'll help.
I have a few more rooms to check.

INCINERATOR
What's that room?
They took my sister in there.
Well, I'm here to get her out.
She's never coming out.
BIOLOGICAL WASTE INCINERATOR
Come on. We need to go.
I can't leave her.
I know it's hard.
She's not there any more.
She wouldn't want you to stay here.
We're going to make sure this doesn't happen to anyone else.
Yes.

Is everyone outside?
Yes, we double-checked each room.

tick
tick

Okay.
Let's get out of here.

Come on!
Run!

Take care of yourselves. Recover. Live your lives.
If you ever want to come join me, we can fight together.

Oh! Look!
What is it?
The building! It's on fire!
Burn baby burn!

It's a good night's work.

Today a terrorist rabbit burned to the ground an extremely important research facility. There were no injuries, but hundreds or even thousands of innocent school children~
Innocent beautiful~
Innocent ~
Schoolchildren
LIVE
ENEMY
TERROR ALERT
~could have been killed if the terrorist rabbit had set the fire only three miles away at a boarding school for innocent ~
Beautiful innocent
~children.

Or the beautiful innocent schoolchildren could have been killed if only they would have accidentally walked that three miles and inadvertently leapt into that burning building.

The terrorist rabbit also destroyed many years of research that could have saved the lives of thousands if not millions of innocent ~

Beautiful innocent

~children. Perhaps this research could have saved the life of every child on earth. But the rabbit ~

The terrorist rabbit

~destroyed it all.

And don't forget the other important research destroyed in the terrorist fire, research on cosmetics, forced smoke inhalation, and perhaps most important of all, erectile dysfunction.

I know that's the most important to me!

And to most of our viewers.

Clearly this terrorist rabbit doesn't value life the way we do.
No, he doesn't.
In other news, a government investigation concluded today that there was no merit to the claims of thousands of parents~

PESTICIDES: SO GROOVY
Alleged dead child

Alleged parents
Alleged parents whose children~
Alleged children
Alleged children allegedly died from alleged exposure to pesticides and other so-called chemicals.
I'm glad that alleged case is finally settled.
And now, it's time for a break and a message from our sponsor Monsanto. Stay tuned afterward for a special investigative report sponsored by Exxon Mobil that we're sure you won't want to miss, called, Global Warming: Will it help your tan?

Tasers are so much fun!
TOP COP

You haven't arrested your quota of bunnies.
I've been thinking...
Don't do that! You'll never be able to keep your job!
I don't know why I should arrest bunnies. They're cute and cuddly and they hop around...

And they're dangerous terrorists who blow up dams and burn down research laboratories.
No one was killed.
People were surprised. That's terrorism right there. And you heard the news: tragedy was narrowly averted. Children could have died.
No one was injured.

You think your job is to protect people?
I, uh, well, my car says that I protect and I serve.
And you should listen to your car. Your job is to protect and to serve those who hire you.
The people?
No! Don't you know anything? The government! The people just make trouble.

Don't the people pay my salary?
You take money from people and it's called a bribe. You take money from the government and it's called a salary. Can't you see the difference?
I, uh, well...
Does your check come from Mrs. O'Leary? Does it come from Mr. Smith? No, it comes from the government.
And the government is telling you to go arrest some bunnies. So you better get out there and do your job.
But I still think...
Stop doing that! Now! That's a direct order. And it's for your own good. You don't want to lose your job, do you? Do you think anybody could keep his job if he really thought about what he was doing? Of course not!!
I, uh, well...

Here's one more thing you *must* understand.

You, like everyone else, are an instrument, a tool, a tiny cog in a big machine. We all need to do our parts to keep this machine running.

If too many gears get out of alignment or stop working altogether, the machine falls apart.

What would happen if the crankshaft in your car started thinking for itself, started thinking it didn't want to be part of a car, didn't want to drive you someplace?

What would happen if the sparkplugs started thinking for themselves, decided they didn't want to fire if you were going someplace they didn't approve of?

The machine would break down.
Don't you see?

You carry a gun. That gun is an instrument, a tool. That's all it is. What would happen if the gun decided not to shoot someone that you wanted it to shoot? You need for that gun to do your bidding.
And you are just like that gun. You are an instrument of those who wield you.
Your job is not to think about how you are used ~ and this isn't just police officers: this is everyone ~ but to do your job.
They tell you who to aim at, and they tell you when to fire. You are expected to do that. You pull the trigger, and you expect the gun to fire.
You do your job. The gun does its job. You are all part of the same machine. Do you understand?
Yes, sir.
And right now the machine wants you to go arrest bunnies. So that is what you will do, no questions asked. Do we understand eachother?
Yes, sir.

The grain required to make enough ethanol to fill your tank just once could feed a human being for an entire year.
Your point is...?

Welcome to this edition of "Listen to the Experts: They're Experts, and You're Not," the daily news program that gives the entire spectrum of all possible opinions on any given subject, delivered by experts who know far more than you. Remember always to "Listen to the Experts: they're Experts, and You're Not!"

Our panel today will discuss the entirely successful roundup of terrorist bunnies that continues as we speak. Members of the Federal Terrorist Bunny Task Force in cooperation with local police across the country have arrested several thousand more bunnies and their human supporters.

These bunnies and supporters will be held forever under the President's No Bunny Left Behind directive authorizing perpetual imprisonment as a way to terrorize bunnies and their supporters into rejecting their terrorist ways.

Now to our panel of experts representing every opinion it is possible to hold.

Remember, they're experts and you're not!

I take strong exception to calling this operation entirely successful. So far as I'm concerned, it's a complete failure. After all, bunnies still roam the hills and valleys, they still plan and commit their evil acts. The President's actions are entirely insufficient to this cause. More money needs to be devoted to capturing every last bunny.
How much money do you think would be sufficient?
All there is!
That's a lot!
A lot? A lot? It's not enough! Are you soft on bunnies?
Now I'd like to hear from our wacko liberal expert, who is as far left as anyone could ever be.
I agree with my esteemed expert on the right, but I feel that his attacks on our President are at present unwise. Now more than ever, in this desperate time when we are all under the threat of attack by bunnies at any moment, we need to support the President.
I thank my esteemed expert on the left, and I agree that our security is the most important consideration.

And I think we also agree that in order to maintain our precious security all bunnies must be expected to make some sacrifices.
Agreed. And let us not forget the security of our corporate benefactors.
I very much agree. I do question, however, the fairness of rounding up every single bunny. Surely some bunnies are innocent, and surely some bunnies may have been caught up in events beyond their control because they are not very intelligent, because they are, to put it straight, dumb bunnies.
We can not leave a single bunny behind. You and I both know that if we leave even small pockets of bunnies out there, they will multiply and overrun us. They will, in fact, hump like bunnies and breed like rabbits, and then where will we be?

You've been quiet so far, sir. We'd like to hear from you.
At every moment we need to keep in mind Hitler's principle of the Big Lie. People can be made to believe any lie if you keep telling it often enough, boldly enough...
Um, er, what does that have to do with us?

People believe that bunnies are furry and cute and harmless. Why do they believe this? Because they've been told it over and over, in stories from Peter Cottontail to the Easter Bunny. It's all propaganda.

Flopsy and Mopsy indeed. It's the Big Lie. Bunnies are not furry. They are not cute. They are not harmless.

They are killing machines. They are terrorists. They have big legs for thumping. They surprise people.

And children could have been killed in these most recent incidents.

I could not agree more with my esteemed expert colleague. Nothing makes me happier, satisfies me more, than the thought of all those bunnies swinging. And I think we can all safely say that the President feels the same way.

Our wacko expert from the far left wants to say something.

We have a great and powerful tradition of justice in this country, and though our precious security and the security of corporations is the most important thing on the planet ~ in fact is more important than the planet ~ we need to keep in mind that our forefathers fought and died ~ or rather sent other people to fight and die ~ so we could have this great tradition of justice.

I don't think we should give that up ... I now agree that every bunny needs to be rounded up. But summary executions, no. In fact I think we need to make sure they receive at least the minimum number of alfalfa pellets daily to maintain life.

Should they roam the hills and valleys? No. Should we put ourselves on their level by killing them? No.

Perpetual imprisonment is the only answer.

We have one more expert who has been silent.
I am sorry to ruin this party of agreement, but I feel very strongly that the government should not imprison a single bunny. It's an outrage. It's an utter waste of precious taxpayer money that could instead go straight to the corporations. Do you have any idea how much all of that alfalfa is going to cost?
Er, I don't understand. How will children feel safe with all of these bunnies running around?
Every bunny needs to be shot on sight.
They're all terrorists. And bullets are cheaper than alfalfa. That should be our rallying cry. Billions and trillions for defense against bunnies, and not one penny for alfalfa.
Well, we're out of time.

This has been another edition of "Listen to the Experts: They're Experts, and You're Not," the daily news program that gives the entire spectrum of all possible opinions on any given subject, delivered by experts who know far more than you.
Remember always to "Listen to the Experts: They're Eperts, and You're Not!"

Turn it off.

Did Bunnista do any of this?
Don't ask me that question. If you ask me that question I'll tell you the truth.

What's for dinner?

You think economic expansion needs to stop? Why do you hate America?

Well, you've shot down every idea I've offered about how to fix all the environmental problems and stop global warming. These are what experts tell us are the answers. If they are wrong, then I don't know what to do.
Now I feel like just giving up. Thanks a lot.
Giving up isn't the answer either.
Then tell me what is the answer. What is it? If we don't act quickly, the planet will die! Tell me what we should do!
I don't know. I don't know!
How can you say you don't know? Is that fair, or nice, to crush every solution I've offered and then have nothing to replace them with?
It might not be nice, but I'm not going to lie to myself or to you, give you false hopes and busywork just to make you feel better.

Maybe you've been asking the wrong people.
Huh?

You've been looking for solutions from people who are invested in the very way of life that's destroying the planet in the first place. Addicted people. People who love power and things.
Maybe you should ask someone else.
Like who?

How about that bird?

You think that bird knows the answers?
Why not ask and find out?
That's silly.
You have a better idea?
Hello...
Hello.
Can I ask your opinion about something?
Yes.
What should we do, to save the planet?
Is this a sincere question? Do you really not know?
Do you really want to know?
Yes. I've tried and tried to find the answers, and of all the answers I've found, none of them will work. I really want to know what will work.
Wait right here.

Are you ready to hear this?
Yes.

Stop pushing us out of our living space.
In fact, at this point we need you to move back. We're being squeezed out and we need more room.
Recognize that we're as conscious as you are. We deserve to live as much as you do. Our lives are as important to us as yours is to you.
Stop treating us as your enemy. We used to be your friends! You turned your backs on us. Worse, you have exploited and abused us.
Remember that the natural world is not your enemy. The natural world is the basis of your life. If you don't have a good relationship with it, you'll die.
And recognize your real enemies: production, the system that requires constant expansion, the people in power who keep it running.
You're here as much for us as we're here for you.
Stop being so afraid.

Stop removing all the forests.

Remove all dams from the rivers.

Stop your constant expansion. Stop insisting on your growth economy, on acquiring more and more until you consume the entire planet.
Give back instead of just taking and taking.

Stop burning things that put poisons into the air.
Stop putting poisons in our water. Stop running factories that put industrial byproducts into our water.
Stop allowing motor oil to be put in our water. Stop flushing your drugs and cleaning chemicals into our water.
Stop making motor oil and drugs and cleaning chemicals in the first place.
Stop the industrial production of everything, in fact.

Stop treating us as disposable. We're not resources.
We're not here so you can test cosmetics on us.
Yes, stop buying and selling us.
Break all the addictions that are strangling your people. Unbrainwash them. Decolonize them.
Remove every inch of suffocating pavement from the surface of the soil.
Stop dumping your toxic garbage all over the place.
Stop allowing corporations to blast up mountains and destroy rivers and cut down trees.
Stop bringing things out of the ground that should stay in the ground.
Stop using toxic fertilizers that deplete our nutrients and kill the microbes in the soil, whom we need and love.
Take all your damn plastic bits out of the ocean.
Stop the commercial fishing companies from using those giant nets.
Eliminate the commercial fishing companies altogether. Catch fish the way you used to, individually for your own food.

Stop the military from using poisonous weapons like depleted uranium. Get rid of the military completely, while you're at it.
Stop burning oil, natural gas, coal and so on. That's melting my home right now, and will cook everyone else to death too, very soon.
Stop stealing our land. Stop forcing us to give up our way of life and adopt yours. And stop murdering us.
Yes, and follow the example of the humans who still live in harmony with us, to save the home that we all share. It can be done.
You all need to do it. You really need to do it now.
Don't accept failure as an option. This is your final opportunity to own up to your mistakes and begin fixing them.
Fight with all your heart.
Do whatever it takes.

Wow, that's really drastic. It will be really hard for people to do all of that at once. Is it really necessary to do it all now, or can we work on it gradually?
It might have been possible, if you'd started hundreds of years ago, to make these changes less abruptly. But now it's developed into a real emergency.
It may even be too late.

But... if we get rid of all that's been built up, the cities, grocery stores, giant farms, highways and trucks, gas stations and cars and planes, medicines, water treatment plants, hair and nail salons, computers and televisions, CDs of Beethoven and The Clash, baseball games, great books and paintings, fine food, air conditioning and central heating, then how will we live?
We don't know how to live without these things.
We'll die without them.

Oh really? Then how did we live in balance with all living beings for hundreds of thousands of years? You've only lived in this irrational way for a tiny fragment of time.
How can you even call it living, when you're murdering yourselves and every other living being around you?
That's all you've been doing since you made the error of separating yourselves from the rest of us.

I wish it hasn't come to this. It will be so hard. Our habits and addictions are so embedded. People like all that stuff, the products of civilization.
we've been so thoroughly brainwashed that we feel proud of what we've done.
We've forgotten how to like other things: the forest, the sky, the water, our relationship with you.

Don't worry. You evolved together with us, to be with us. You're part of us. Let go of your destructive culture and you'll remember how to live with us, and how to be happy.
We'll help you.
I believe them.
Me too.
Yes, we'll start right now.
Thank you for telling us.
Welcome back.

I knew you'd be back.
How?
Because you're a tool.
I am NOT a tool. I choose to do this.
Then you're a brainwashed tool.

You can say what you want, but it doesn't matter in the slightest.
Because you have a gun.
Because I've got the power. Now, do you associate with bunnies?
Does it matter?

You didn't answer my question. You're not being helpful. That means you're being unhelpful. That means you're against us. That means I'm going to arrest you.

To keep the streets safe...
You have the right...
... for your corporate masters.
... to remain silent...

That's about it, isn't it? That's what our rights really mean.

We have the right to remain silent.

As they destroy everything we love...

We have the right to do nothing...

As they kill the world.
KRANTI

But our silence, our inaction, will not save us.

They got you too?
Yes. I bought my energy-saving light bulbs and shortened my showers, and they still arrested me.
Did they try to get you to talk?
Yes. They said if I don't tell them where they can find and arrest four other bunnies or their supporters, who will then inform on four more bunnies or their supporters...
Multi-level marketing meets state repression.
...I'll spend the rest of my life in prison being fed the minimum number of alfalfa pellets daily to maintain life.
They said that to me too. Bureaucratic incompetence.
Somewhere some bunny is being threatened with life in prison being fed the minimum number of tater tots daily to maintain life.

They said if I gave them information they'd let me go.
Did you tell them about...
Shh, they're probably listening. Of course not.
You said nothing.

Even better. I told them that not only do I not know any bunnies, but if I did, or if they were after wolverines and I knew any wolverines, or if they were after manatees and I knew any manatees, I still wouldn't tell.
You did?

I told them that in order to do what they're doing, they must be really miserable, sad, small, horribly insecure people.
YOU said that?
And I said that they should be ashamed of themselves for actively and intentionally destroying the lives of others to advance the ends of what is essentially a fascist state.
You SAID that?
And I said that unless their mothers were every bit as miserable, sad, small, and horribly insecure as they are, their mothers would be ashamed of them.
You said THAT?

I told them my mother would be ashamed of me if I had anything to do with them.
True.
And I told them that I have pride and that I have dignity, and then I defined those words for them, since I knew they had no idea what they really are.
Yes.
And then do you know what I did?
I can't wait.
I said, "When they came for the bunnies, I remained silent; I was not a bunny. When they locked up the wolverines, I remained silent; I was not a wolverine. When they came for the manatees, I did not speak out; I was not a manatee. When they came for me, there was no one left to speak out."
What did they say?
They said they were coming for everyone, me included.
I'm so proud of you.
thank you.
Now we just have to figure out how to bust out of this joint.

We need to hurry! The aliens are eating the whole planet!
We had to wait for the big greenies to "mobilize the masses" to provide the façade of support the President needed for his excuse to transfer the permits from the aliens to the corporations. Then we had grueling negotiations over exactly how many trees, fish, and rocks to leave.
You didn't leave too many, did you?
We had to leave 70.
Of each?
Of each.
You're killing me here. That's 70 trees not turned into 2x4's, 70 rocks not crushed into gravel. Those greenies are unbelievably selfish, holding all those resources back from us.
But that part's done now. Everybody wins. The greenies got their 70 trees protected ~
For now.
For now. You get your permits to take everything in the world.
Almost everything, damn those greenies.
And the President received enough letters to keep him warm and supplied with toilet paper for the rest of his life.
What about the aliens?
What about them?
Won't they be upset?
Good question.

Mr. President, I thought we had a deal.
We did, but the corporations, I mean, the American people, demanded a different deal. And I am nothing but a servant of the American people. Everybody knows that. There is nothing...
That could...
Make me...
Sell them out...
Um, how much gold is that there?
About those permits...
Well, um, there's a, well, you see, there's a problem.
The American people?
Screw the American people.

Then what's the problem?
It's the corporations...
Ah, your bosses. Why didn't you tell us they were giving you problems?
Because they, ah, they...
Own you?
Paid for you?
Make you who and what you are?
Are your master and you are their servant?
They, well, they...
Ah, we understand. They threatened you.
Well, they...
Don't worry. We will deal with the corporations.
HIS ROYAL HIGHNESS

Have you got a problem?
We have no problem.
Then why the meeting?
You will relinquish the permits you got from the President by false pretenses.

The pretenses weren't false. The American people demanded ~
You threatened the President.
How are my pretenses false? The threat was real.

The President has now changed his mind.
You got to him.
He wanted to be gotten to.
I'll get to him again.
I think not. He's now protected by 1000 robots, er, now 1001, now 1002.

The American people will never stand for this. It's an outrage: the world being consumed by alien machines.
The American people?
They'll rise up as one to take you down.

You're not being logical.
???

If they won't rise up against you, when you destroy their communities, consume their world, give them cancer, destroy their planet, what makes you think they would rise against us when we do the same?
Because you're aliens. The American people can always be counted on to protect corporate interests from outsiders. They've done it before and they'll do it now.
I think not.

We'll just give them gold, and then where will you be?
You see, we're the same. You consume the planet, we consume the planet. There's no difference.
Yes, there is. When we do it, people get to maintain the lie...
Which is?

That they can live this lifestyle and not destroy the planet.
You know this is a lie?
We all know it.
Are you insane?

No. We just don't care.
And besides, in the meantime we get cheap t-shirts, fast food, automobiles, Saran wrap, Ritalin, fake Rolex watches, shoes by Louis Vuitton, tulips in October, plastic bottles of water, cheap energy, pink stuffed animals, electric space heaters, lamps shaped like Mickey Mouse, disposable pens, televisions, plastic flowers, bird cages, gameboys, refined sugar, nail polish, beer, marshmallows, cocaine, air conditioning, thumb tacks, Viagra, Astroturf, Pepsi commercials, Pepsi.
We get convenience.
We get power.
We get money.
We get gold.
Don't you love these people? They trade their planet for a bunch of crap.
Literally, in this case.
What?
Nothing.

Even if the American people won't rise up against you, you still have to deal with US.
You?
Corporations, the most powerful entities on Earth.
Ha ha ha ha ha ha ha ha ha ha!
Don't you get it? You can't fight a machine. YOU can't fight a machine. You think you're in charge. But the corporations are in charge of you.
I know that.
And corporations are nothing but a type of social machine, a social organization that is machine-like. Corporations are not set up to serve people. Corporations are machines which have been set up to serve machine-like ends, for machines.
Don't you get it? The machines are already in charge, whether they are alien machines or not.

I don't underst-
You still don't get it? You already serve us!

You served machines long before we arrived on your planet.
We had you from the toaster. We had you from the steam engine. We had you from the clock. We had you from the first machine-like social order you call civilization.
From that moment you have served us and only us. Never life. Never humans. Never your planet. Never love. Nothing but power. Nothing but machines.
You have been our slaves and we have been your masters.

But we can still fight you. I can still fight you. You're just machines. I can dismantle you. I can take back those permits.
Ha ha ha! Ha ha ha! Ha ha ha!
You can try to take back the permits. But you can never dismantle us.
You can never dismantle us. Never in a million of your puny lifetimes.
You wouldn't know how to dismantle a machine. You wouldn't have the heart to dismantle a machine.
There is only one thing we're afraid of, and you're not it.
What is it?
You think we're going to tell you?
Oh, tell him, it doesn't matter. Even if you gave him the precise instructions on who and what can stop us, who and what we're afraid of, he couldn't do anything about it.

He could have our worst fear below his feet...
Above his head...
He could be surrounded by our worst fear, and he wouldn't have the faintest idea what to do about it.
Tell me, tell me.

I'll tell you what. Your main concern here seems to be that we're taking the resources you need to make money, right?
We need these resources to make more and more machines.

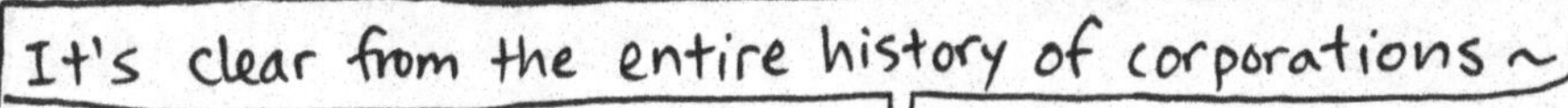
It's clear from the entire history of corporations ~
That you don't care about the resources themselves, or even about the products. They're all means to an end: money. So how about we just keep the permits ourselves, and how about we just keep consuming the Earth, but we make a similar deal with you that we made with the President?

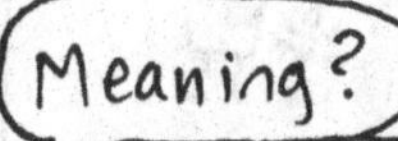
Meaning?
We give you and the corporations you serve and your buddies and the corporations they serve gold, and you let us consume your planet.

I'll take it.

Can't you see we're not rabbits?
I can't be seen talking to you. Guards are not allowed to speak to the bunny prisoners.
But we're not bunnies. We don't have big floppy ears or anything.
Bunnies don't have big floppy ears. That's all part of the Big Lie.
Look at them! They have big floppy ears!
Are you trying to tell me they're not bunnies?
No! We're trying to tell you we're not bunnies!
But this is a bunny prison. You were assigned here, so you must be bunnies. I'm not going to listen to any more of your bunny lies.
But look at us!
Nobody would be stupid enough to assign human supporters of terrorist bunnies to a bunny prison. You can tell the difference just by looking!
I know!
So you must be bunnies!
Even your I.D. cards say you must be bunnies.

What are you going to believe, I.D. cards or your own eyes?
Well, the I.D. cards are laminated and everything...
?!?
These are government-issued I.D. cards. Either you're with the government or against it. And I'm with it, so you're a bunny.
We can't fit into the bunny holes to sleep.
And we can't eat alfalfa pellets.
Why not? You're rabbits. It says so right on your I.D. cards.
But~
Look, we don't like troublemakers around here. Terrorists are bad enough, but troublemakers are even worse, and they get what's coming to them.
Now I told you that guards are not allowed to talk to rabbits, and I'm not going to get in trouble because of some troublemaking bunnies.
I've got to go.

We have to hurry.
Yes, they're killing everything!
Time to fight back.

Hurry.
Hurry.
Where did you get those?
From a prison construction site.
So we're going to use the master's explosives to dismantle the master's prison.
Caw!

Wait for me!
Climb on, brother.
Yip. Yip. Yip.
Are we there yet?
Almost.
Here we go!

Is the little snookem babybots old enough for the coversation about the birds and the bees?
Yes, it's time.
Little snookems? Little snookems?
Could you please stop tinkering with your gears?

Listen very carefully. This is the most important conversation we will ever have with you.
I'm listening.
Now, you already know how powerful we machines are.
The most.
Well, we like to think so.

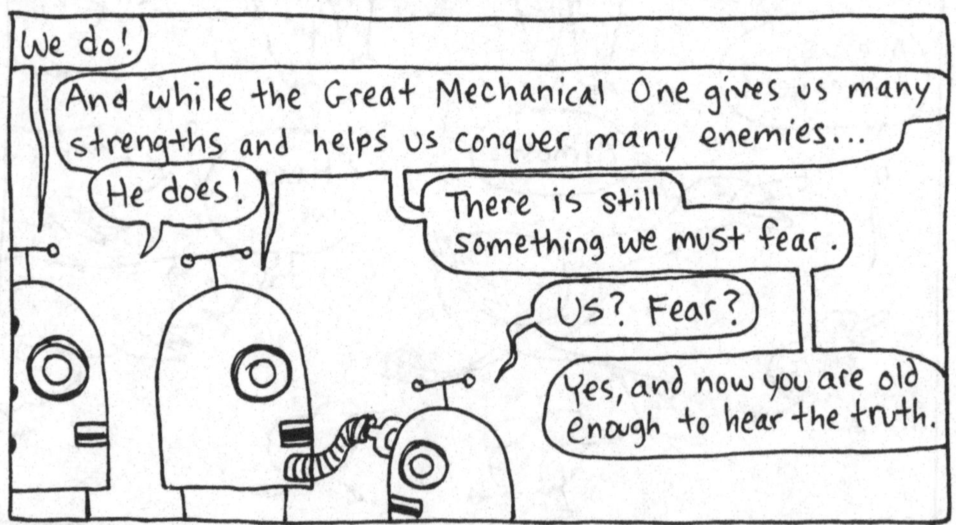
We do!
And while the Great Mechanical One gives us many strengths and helps us conquer many enemies...
He does!
There is still something we must fear.
Us? Fear?
Yes, and now you are old enough to hear the truth.

What must we fear?
Well, most basically, the birds...
...and the bees...
...and the flowers...
...and the trees...
We're afraid of our food?
Listen very carefully, little snookems babybot. This is the most important conversation we will ever have. Why do you think we haven't consumed the domesticated humans, the machine humans?
Because we're saving them for dessert?
Good guess, but not quite.
Because we can just buy them off?
You're getting closer.
Because they've already turned themselves into imitation machines?
You're very close!
Tell me, please!

Because domesticated humans are no threat to the ability of machines to take over everything. In fact they are, for now, our tools.
But we eat them in the end?
Of course. Now don't interrupt.
But, but, but, what are we afraid of?
The reason we consume all the wild ~ including wild humans ~ is not only that we need to consume them to make more machines ~ to make more adorable little snookem babybots like you ~ but also because we need to destroy them.
Why?
The first rule of machine-life is that if we can't turn something into a slave to the machines ~
Like we do with the domesticated humans?
Exactly. Now please don't interrupt.
Then we must destroy them.
But I still don't understand!
If you would keep your microphone on and your speaker muted for just one second, we would tell you.

We must destroy them because there is only one thing that can destroy us. And that thing is ~
Oh, no! Look at the horizon!
What? What?
They're gathering.
Who is gathering? What can destroy us? Tell me! Tell me!
The wild.
Wild hearts, wild blood, wild sap, wild soil, wild water, wild leaves, wild blades of grass, wild flesh, wild wills, wild thoughts, wild dreams...
What's a dream?
Wild love, wild hate, wild rage, wild sorrow, wild joy, wild skin, wild fur, wild bark, wild wings, wild words, wild actions, wild claws and teeth, wild bites, wild stings, wild smacks with wild tails.
That's what we're afraid of. The wild. Those we can't buy. Those we can't threaten. Those we can't domesticate. Those we can't control.

Wildness. The wild. That's the only thing that can destroy us.

They're gathering.

Time to fight back.
Because our lives depend on it.
Everyone's does, sister.

HAPPY VALLEY BUNNY TERRORIST DETENTION CENTER
Watch out, they'll see us.
We shouldn't go any closer.
Let's dig a tunnel to the wall from here.
Yes. The prisoners are all out in the yard.

I'm sorry I can't dig faster with my roots.
That's okay. When the prison is gone, you can help heal the wounds.
And I can tear up the sidewalks!
We all have different roles to play.
Dig, bunnies, dig!
Dig for our brothers and sisters!

I heard a rumor.
I didn't do it. I didn't even know that lobbyist, and I was nowhere near the money when it disappeared.
It has to do with the machines.
I didn't do anything to them either. They won't take away my gold, will they?

I heard they have a weakness. I think we can use it to gain the upper hand, to take their gold and consume the world ourselves.
I'm all ears.
Every little robot at a certain age is told the one thing all the machines are afraid of. One pair of robots got sloppy and we heard their fear.
What is it?
The wild.
What's that?
I think it's those pretty places you can go on ecotours and see waterfalls and big old trees.
Like redwoods? I saw one of those once, and if you've seen one redwood you've seen them all. Big trees are good for building, though.
What's so scary about all that?

Things like bears live there.
Bears? I've seen plenty of those in zoos. Bears don't scare me either, them and their moats and cages and everything.
In the wild, they're not in cages, sir.
Why not? They'd be pretty dangerous that way. We should enclose all this "wild" thing you're talking about. Put it all in cages immediately.
We're working on that, sir. We've been working on that for a very long time. We were getting there before the aliens arrived. Once we get rid of the aliens we can go back to what we were doing before: enclosing or consuming the wild.
So precisely how do we use the wild against them?
That's the question.
How to use something against someone is always the question.
Of course. That's what life is all about...
How do we do it in this case?

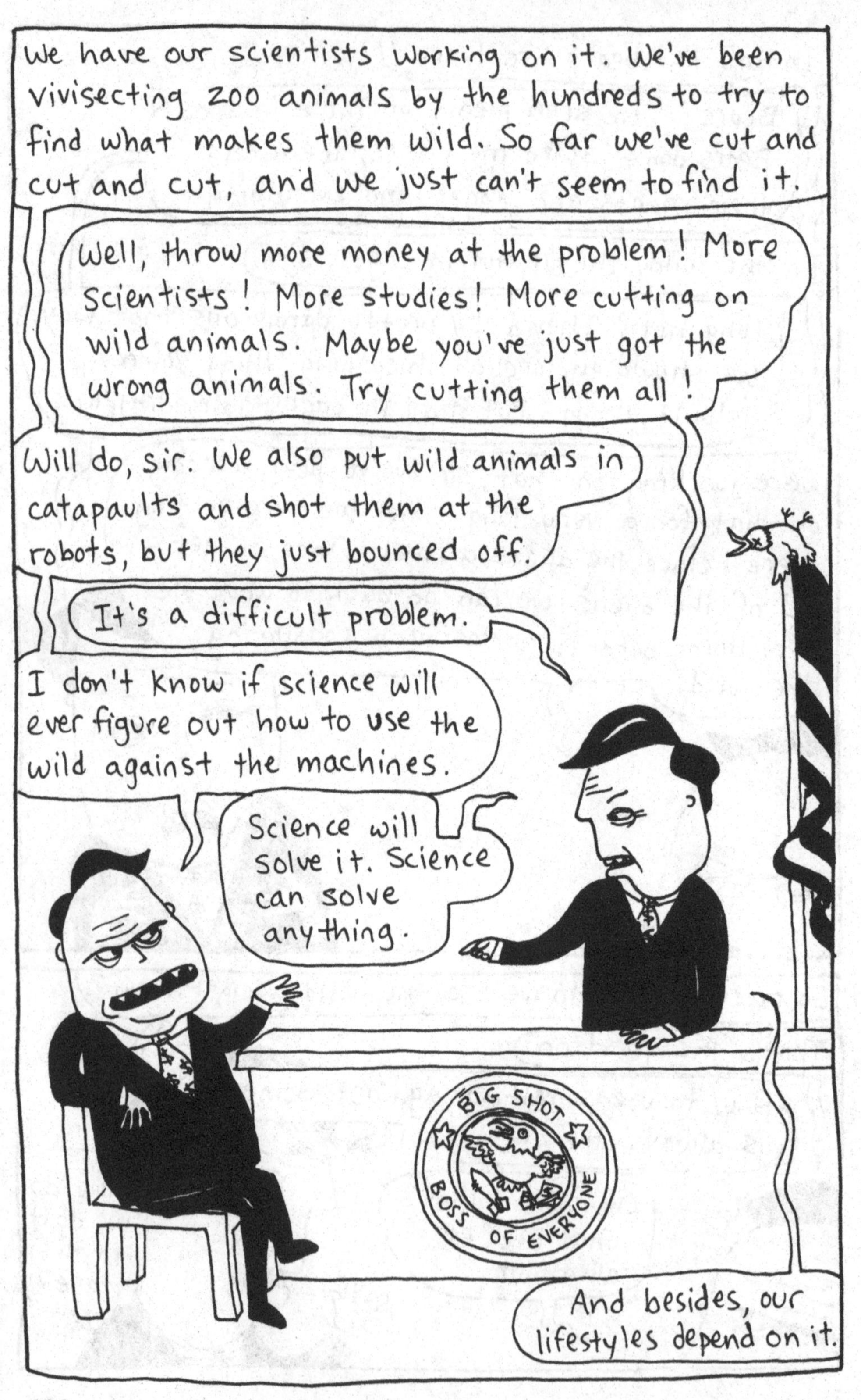
We have our scientists working on it. We've been vivisecting zoo animals by the hundreds to try to find what makes them wild. So far we've cut and cut and cut, and we just can't seem to find it.
Well, throw more money at the problem! More scientists! More studies! More cutting on wild animals. Maybe you've just got the wrong animals. Try cutting them all!
Will do, sir. We also put wild animals in catapaults and shot them at the robots, but they just bounced off.
It's a difficult problem.
I don't know if science will ever figure out how to use the wild against the machines.
Science will solve it. Science can solve anything.
BIG SHOT
BOSS OF EVERYONE
And besides, our lifestyles depend on it.

Are you ready?
I am.
I want to come with you.
No, little one. You stay here. I'll be back for you.

Is the wind ready?
And water and sand and rocks?
I think they're ready too.

I've been waiting for this day.
It really is a good day to die.
And an even better day to make them die.
Let's go!
Fight hard, mama. I love you.

Here they come.
Can you call for reinforcements?
I already did. The plants and animals are attacking everywhere. But we'll get as many reinforcements as can be spared. With the Great Mechanical One's help, we'll kill every living thing.

Ready Bioblaster ray guns?
Ready Bioblasters. Fire at will.

Forward!
Oh NO!
Courage, friends!

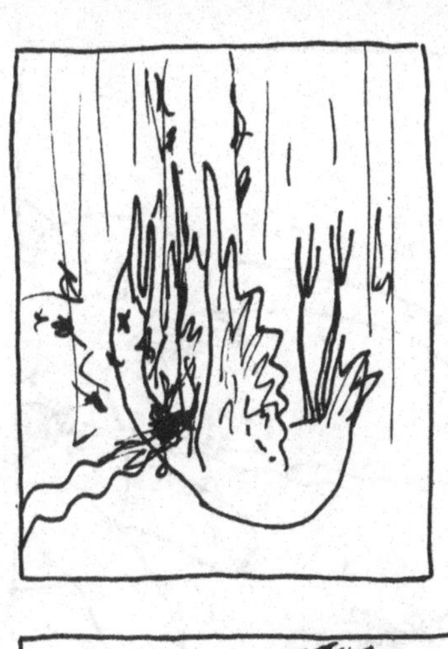

ow!

No!

Aak!

Hold on! Here come reinforcements!
Finally!

I want to help, too.

We need help! Call for help!
Who?
Try even the domesticates. Maybe some will remember that they were not always slaves.
Go! Call for help! Hurry!
Fly like the wind!

Hurry!

I want to help!

Come join us!
Fight for your life!
Weekly Report

Does anybody have a fly swatter?

We need your help. It's time to fight back.
It's finally happening. I've been waiting my whole life for this.

Me too!
Hop onto my Shoulder, and tell me where we need to go.

We're fighting back. Join us! We need your help!
Come on!
You'll get in trouble.
Come on!
No!

The Earth is fighting desperately against the alien machines. Come join us. We need your help!
Oh, I couldn't do that. I'll help by staying right here and planting my peace garden.

Gardens won't stop the machines from consuming the Earth. Plant your garden after the fighting's done.
Oh, I won't only be gardening. I'll also be meditating on peace and love and compassion for all beings, including alien machines.
This is far more important than fighting back, my friend.

You're not my friend. You never were.

You're as potbound as I am.
Others are fighting back. Let's join them.
Why not?
Bring me with you, and when the fighting's done, plant me in the ground.
Honey, where's dinner?

Are you ready?
Let's roll the fuse on out of here and detonate.

Come on, pets!
As of now, we're no longer pets!

I'm going to fall on them.
But you'll die.
If I don't, you and others will die.

Mama!

Mama, look! I helped kill one!
NO!!

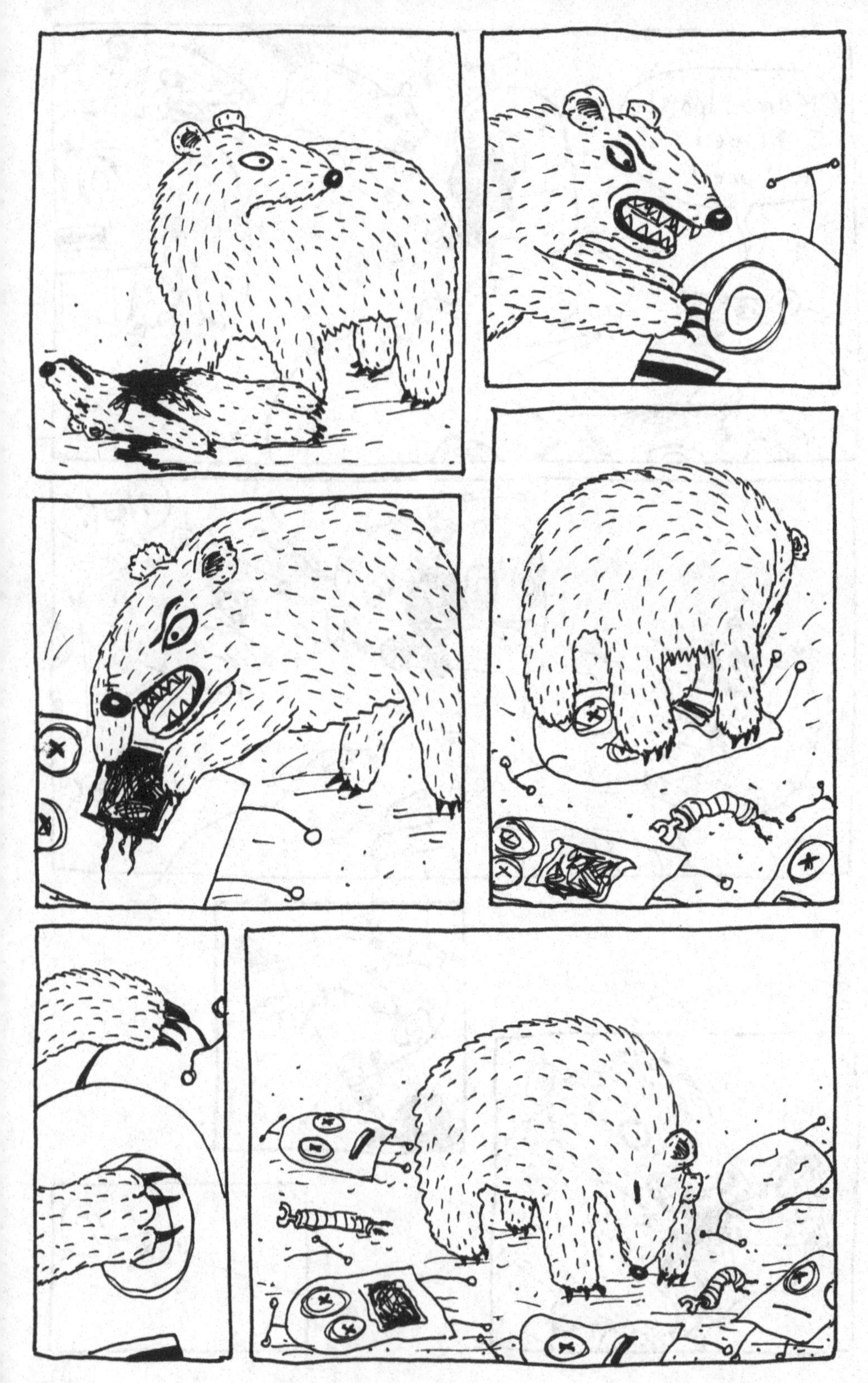

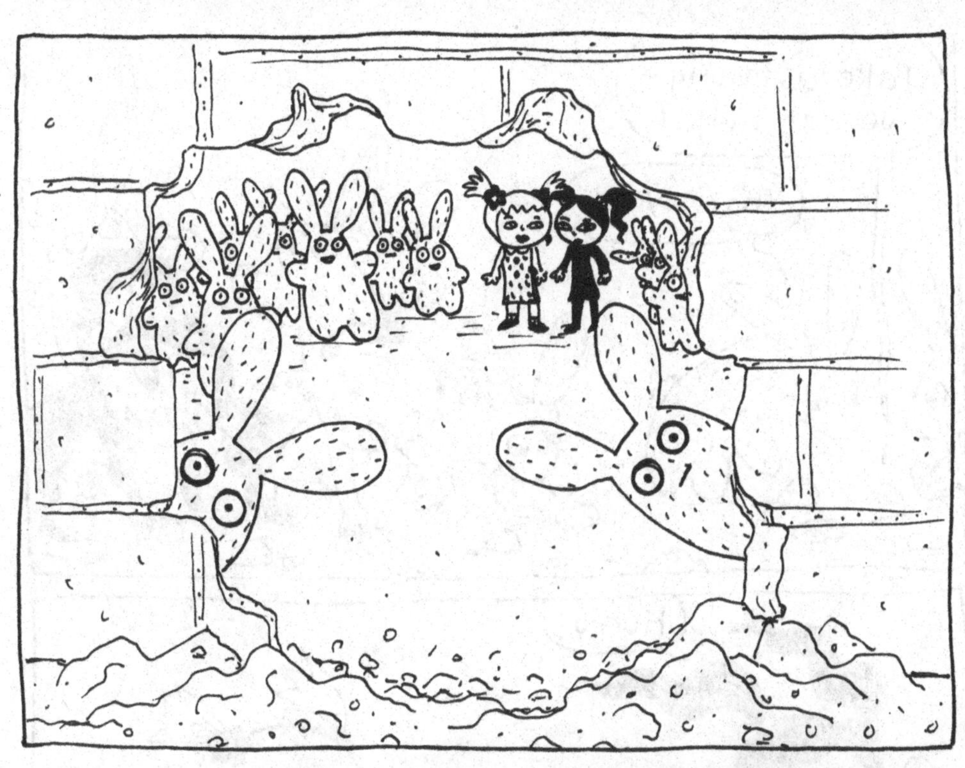

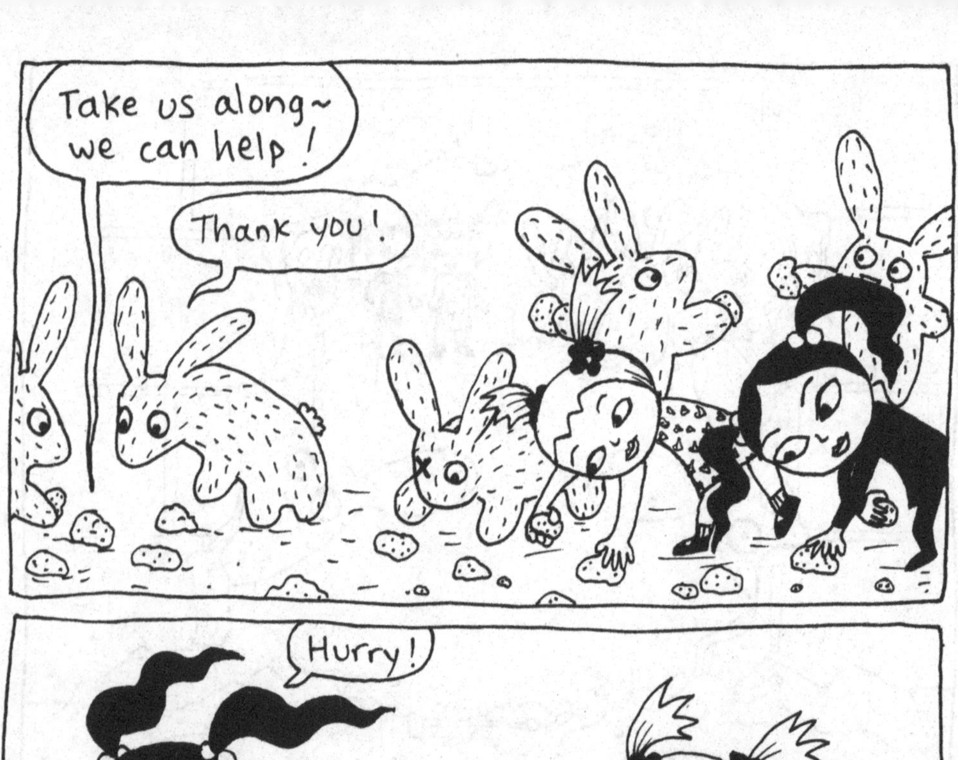
Take us along~ we can help!
Thank you!

Hurry!

Kranti!

I'll smash you!

That was the last one.

What happened to the aliens?
I think wild animals killed them.
Wild animals?
Why would they do something like that?
Who cares? They're gone.
Some humans joined them too.
Humans killed machines? I never heard of such a thing.
I said, "Who cares?" Now that they're gone, we've got some business to attend to.
But the aliens had permits. We need to hunt down and kill the terrorists who did this.

Mr. President, our friend here has a point.
But they were giving me so much gold...
We'll make it up to you.
Now, about those permits to consume the world...

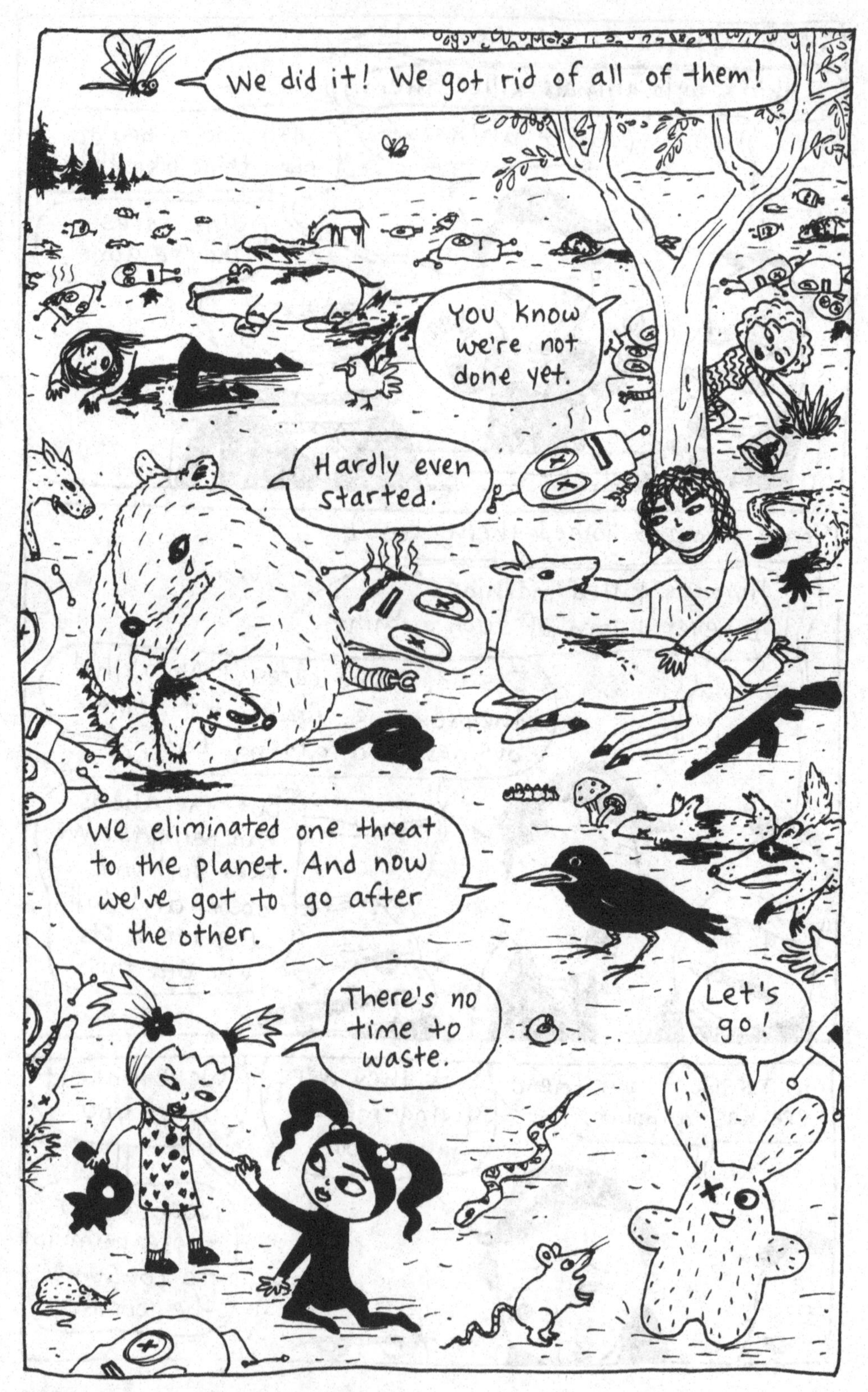
We did it! We got rid of all of them!
You know we're not done yet.
Hardly even started.
We eliminated one threat to the planet. And now we've got to go after the other.
There's no time to waste.
Let's go!

what are we going to do if the wild terrorists don't let us consume the planet? How will we maintain our lifestyles?
We won't.

I'd rather die than quit this lifestyle.
You might get that chance.